Food, farming and the countryside: past, present and future

An account of the part played by food and agriculture in the evolution of everyday life and of the English countryside

CONTENTS

Acknowledgements	vii
Chronology of food and agriculture	ix
Preface	xxv
1 Agriculture, food and everyday life in Britain	1
2 From hunter gatherer to farmer	55
3 The intensive livestock industries	81
4 Agriculture, food and the countryside: past, present and future	109
Epilogue	151
Index	155

ACKNOWLEDGEMENTS

I should like to thank the following people and organisations who have made the preparation of this book possible:

Firstly the editorial inputs by Nottingham University Press were considerable, and both the structure and the content of the book reflect their influence substantially.

Much of the source material was researched in the James Cameron-Gifford Library of Agricultural and Food Sciences at the University of Nottingham, Sutton Bonington Campus. Particular thanks are due for permission to consult the rare books collection and the archives.

Another library used profusely was the Hallward Library of the University of Nottingham. I am grateful for access to the resources of the Department of Manuscripts and Special Collections, University of Nottingham, and for permission to reproduce examples from their extensive photographic collection.

Photographic studio work was by Laura Allard of the Photovisual Unit at Sutton Bonington Campus.

Over the years many ADAS and University staff, and John Cessford of BOCM PAULS, have influenced or inspired aspects of the project, usually unwittingly.

Peter Lewis, Sue Gordon, David Edgar and Jim Holton made helpful comments or provided source material. William Wood and John Orme donated historical textbooks to my collection.

Many people and organisations have generously provided photographs or permission to use them. The library of the Royal Agricultural Society of England (librarian Phillip Sheppy) permitted me to reproduce the distinguished portrait of Robert Bakewell. Historically important slides from ADAS poultry archives have been used with the help and permission of Andrew Walker. The National Federation of Women's Institutes allowed me to reproduce some of their historic photographs. Thanks are due to the Pile family for the photos of the farm shop at Middle Farm, Firle, Sussex, (photos by John Pile), and for hosting the book launch.

Suzie Matthews and colleagues of the Countryside Agency supplied the farmers' market photograph. Tesco provided pictures of modern food retailing. The photograph of a jungle fowl is by John Corder of the World Pheasant Association. The picture of the Staffordshire dairy instructor in the 1920s was donated by Marjorie Orme. Several pig and pig product photos are by Des Cole and John Gadd. Des Cole also supplied some food retailing pictures. Sarah Keeling provided a countryside picture. F. Bailey & Son, family butchers, of Upper Broughton, Melton Mowbray, provided a picture of their pork pies.

Photographs of Keyworth old and new were compiled by Nigel Morley of Keyworth and District Local History Society. The picture of a modern dairy cow was by Adam Lock.

The Trustees of the Laxton Visitor Centre permitted publication of their plan of the Laxton Estate.

Reproduction of Pierce's 1635 map of Laxton and agricultural scenes is by permission of the Department of Special Collections and Western Manuscripts, Bodleian Library, University of Oxford.

Some of the author's photographs were taken with the permission of the owners or managers of the locations. These included: ADAS Gleadthorpe; Holkham Estate, Norfolk; Farm and Country Centre, William Scott Abbot Trust, at Sacrewell, Peterborough; Eglantine Vineyard, at Costock, Nottinghamshire; Merryweather's garden centre and Bramley Apple Museum at Southwell; and Naturescape Wild Flower Centre at Langar, Nottinghamshire. Leicester City Council, Leicester Museums, (managing curator Stewart Warburton), permitted publication of the photo of the 1937 mobile fish and chip shop.

CHRONOLOGY OF FOOD AND AGRICULTURE

70 million years ago	Primates emerged; herbivorous diet
20 million years ago	Scavenging, hunting and gathering started by ape-man higher primates
2 million years ago	Tools used for preparing food. Cool storage in caves
1 million years ago	Butchery tools used
800,000 years ago	Use of fire made cooking possible
100,000 to 200,000 years ago	*Homo sapiens* emerged
30,000 BC	Invention of the hole and stone oven by *H.sapiens*
12,000 BC	Food preservation by drying developed
10,000 BC	Honey gathered
10,000 BC	Wild grapes gathered
9,000 BC	Domestication of the sheep
9,000 BC	Cultivation of Einkorn wheat
9,000 BC	Salt traded commercially
7,000 BC	Domestication of the goat
7,000 BC	Domestication of the pig
7,000 BC	Cultivation of the bean
6,000 BC	Domestication of the chicken
6,000 BC	Domestication of cattle
5,500 BC	Fermentation first used, Iran
5,500 BC	Olives domesticated
5,000 BC	Chinese farmed rice
5,000 BC	Incas farmed potatoes
4,500 BBC	Fermentation by yeasts (to make bread and wine) discovered
4,000 BC	Cheese invented
4,000 BC	Sheep reached Britain
3,500 BC	Orchard crops appeared
3,000 to 2,700 BC	Domestication of the goose
2,500 BC	First agricultural text book published by Sumerian scribes
2,500 BC	Quern mill
1,000 BC	Cattle reached Britain
1,000 BC	Domestication of the duck

1,000 BC	Invention of the scythe
by 700 BC	Herbs used, lentils used
8th century BC	Hesiod's advice to farmers written
c.585 BC	Thales of Miletus, first agricultural scientist, used weather forecasting to predict the olive crop
5th century BC	Herodotus, first food writer
c.400 to 300 BC	Greek Hippocratic school influences dietary recommendations. Use of liver for night blindness presages later discovery of vitamin A
c.75 BC	Mould board plough reached southern England
1st century AD	Lucius Junius Moderatus Columella, Roman agricultural author, writing his 13 books
1st century AD	Romans bring herbs such as parsley, sage, fennel, thyme and rosemary to Britain
2nd century	Roman physician Galen made dietary recommendations
6th & 7th century *et seq.*?	Open field system developed
c.1000	First sugar refinery at Candia
11th century	Origins of Crown Estates
1124	Poulterers Livery Company formed
1180	Guild of Pepperers mentioned in London, later called Grossarii, or grocers
12th century *et seq.*	Enclosure
13th century	Goose fair established at Nottingham
1304	Guilds of White Bakers and Brown Bakers in London
1324	Royal decree declared that 1 inch = length of three barley corns
1348-1349	Black death hastened demise of the manorial system, some villages abandoned
1440	Early food regulation, London bakers regulated
1495	Marmalade introduced to Britain from Portugal
16th century	Turnips and clover first introduced
16th century	Cheddar cheese first made
1541	The turkey reached England
c.1560	First cook book published
1570	Spaniards brought the potato from the Andes

1577	Barnaby Googe wrote of reaping machines in the Netherlands and on roots and clover in rotations
1594	Sir Hugh Plat recommended roofed manure stores
1594	Sir Richard Weston published observations on rotations from the Netherlands
1596-1625	Sir Henry Hobart of Blickling improved estate management techniques
1600s	Anglo-Dutch disputes over sources of spices
1600s	Potato reached England
1610	Arrival of coffee in Europe
1611	Introduction of the table fork from Italy
1616	Nutmeg island of Run (Indian Ocean) claimed for England by Courthope
1650	Drinking chocolate reached England
c.1650	Dietary recommendations based on Greek four humours changed to Paracelcus's three humours. Blancmanges and sauces gave way to recognisable modern dishes. Sugar scorned so desserts relegated to separate dishes
1651	First commercial patent for food preservation
1657	Chocolate house opened in London
1664	Arrival of coffee in England
1667	British acquisition of Manhattan in exchange for nutmeg island of Run
1674-1738	Turnip Townshend, developer of Norfolk four-course rotation
1674-1740	Jethro Tull, developer of the horse hoe, seed drill and associated husbandry
1689	First of numerous Corn Acts
18th century	Citrus fruits used for curing scurvy
1700-1710	Tea drinking popular in Britain
1718-1792	4th Earl of Sandwich, inventor of the sandwich
1725-1795	Robert Bakewell, livestock improver
1731	Jethro Tull published *New Horse Hoeing Husbandry*
1738-1820	George III (Farmer George), reigned 1760-1820

1741-1820	Arthur Young, first major agricultural journalist
1745	Columella translated into English
c.1750	Widespread systematic enclosure began
1754-1842	Thos. Coke of Holkham, populariser of Norfolk four-course rotation
1763-1835	William Cobbett, journalist and rural commentator. Wrote *Rural rides* (1830).
1765	First restaurant opened in Paris by Boulanger
1770s	Tours by Young and Marshall reported on the state of agriculture
1776	Adam Smith advocated free trade
1777	Lavoisier showed that animals burn oxygen to form carbon dioxide
1781	First formal food and restaurant critics, Paris
1782	Parishes made responsible for employment, by Gilbert's Act
1784	Meikle's threshing machine developed
1784	Arthur Young's *Annals of Agriculture* first published
1785	Highland clearances began
1789	Restaurants popularised in Paris after revolution left no aristocrats for chefs to feed
1790	First chair of agriculture founded at Edinburgh University
1793	Board of Agriculture established by Sinclair to encourage technology and collect statistics
1794	First account of factory farming. More than 11,000 pigs/year fattened on brewery waste in Vauxhall, Battersea and Wandsworth
1795	Board of Agriculture encouraged corners and wastes to be used as allotments
1795	Speenhamland system for parish poor relief
1801	General Enclosure Act, applied to commons
1803	Sir Humphry Davy's lectures on agricultural chemistry
1803-1873	Liebig, whose realisation of the connection between food and heat later led to developments in livestock housing and human dieting
1804	Appert's sterilisation experiments
1809	Bramley apple bred at Southwell, Notts
1810	Steam plough first patented
1810	Durand's canning developments

1812	Appert developed bottling for food preservation
1812	Donkin developed canning
1813-1834	Poor Laws
1815	Lord Liverpool's Corn Laws to protect home industry
1819	Maypole Dairies founded (see Allied Suppliers, 1929)
c.1820	Baked potato carts on London streets: an early fast food
1822	Board of Agriculture dissolved
1822	Coates's *Shorthorn Herdbook*
1825	Cox's orange pippin bred
1828	Amendments to Corn Laws
1828	Patrick Bell's reaper
1830	William Cobbett's *Rural rides* published
1830s	Fried fish shops in London: another early fast food
1831	McCormick's reaper
1833	First trade union formed by Tolpuddle Martyrs
1835	First repeal of the Corn Laws
1836	General Enclosure Act, applied to open fields
1838	Agricultural Society of England founded
1838	Protein discovered by Mulder
1839	First national agricultural show, Oxford
1839	Invention of digestive biscuits in an attempt to prevent flatulence
1840	Royal Agricultural Society of England granted royal charter
1840	*Journal of the Royal Agricultural Society of England* first published
1840	Liebig's *Chemistry in its application to Agriculture and Physiology*
1841	Downes Edwards's dried potatoes
1842	Superphosphate manufactured by Lawes
1842	Farmers' Club founded, some political activity
1843	Rothamsted Experimental Station founded, by Lawes and Gilbert
1843	*Farmer and Stockbreeder* magazine founded
1844	Rochdale Pioneers started co-operative shops, which led to multiple stores
1845	Enclosure Act, dealing with commons as recreational areas

1845	Royal Agricultural College opened
1845-1847	Irish potato famine
1846	Repeal of the Corn Laws
1847	Vegetarian Society founded
1847	Mulder's dietary recommendations
1848	Liebig's meat extracts developed
1849	Marshall's threshing machines introduced
1850s	James Harrison's ice factory
1850s	Potato crisps inadvertently invented by George Crum, Saratoga Springs, NY
1850-1870	Table laying changed from *à la Française* to *à la Russe* (i.e. one course at a time)
1850-1874	Golden age of agriculture, High farming
1851	Reaper first used in England
1856	Standards for farmhouse Cheddar cheese published
1856	Gail Borden's condensed milk patent
1858	Kilner's preservation jars
1859	Earliest refrigeration
1860	Fowler's double engine ploughing tackle
1860	First Act of Parliament to address food purity and safety
c.1860	Fish and chips became popular
1860s	Discovery that proteins were composed of amino acids
1860s	Edward Smith's dietary standards
1861	Pasteur's germ theory of food spoilage
1861	Mrs. Beeton's *Book of Household Management*
1862	*The Grocer* magazine founded
1864	First London tea room opened by the Aerated Bread Company
1865	Gregor Mendel's experiments on peas provide foundations for genetics
1865-1866	Rinderpest (cattle plague) killed 324,000 cattle
1866	Agricultural Returns instituted (annual collection of statistics)
1867	Henri Nestlé's baby food developed, led to world's largest food company
1869	J. Sainsbury opened shop in Drury Lane, London
1869	Heinz founded
1870	First co-operative cheese factory, Derby

c.1870	Margarine developed
1870s	Bird's custard, an early ready prepared food
1870s	Kellogg brothers developed cereal foods for use in a sanatorium, Michigan
1870s	Fish and chip shops became popular
1870s	Pigs kept in London boroughs of North Kensington and Southwark caused complaints
1871	Lipton's opened
1871	Bamford's farm implements started
1872	Agricultural labourers' trade union founded by Joseph Arch
1872	Local authorities and public analysts given powers over food safety and purity
1874	Free trade brings depression in prices of agricultural products
1874 to 1890s	Widespread switch to dairying after N.American grain made cereals less competitive
1875	Food and Drugs Act tackled adulteration
1875	Milk chocolate developed by Daniel Peter, Vevey, Switzerland
1877	Laval's cream separator
1879	Reaper with knotter contributed to cereal mass production in New World
1880s	Factory baking began
1880s	Factory made foods included biscuits, jam and chocolate
1880s	Glass house cropping and commercial tomatoes developed
1881	Hearson's incubator patented
1883	The *Agricultural notebook* first compiled by Primrose McConnell
1884	Lyon's tea shops popularised eating out
1884	Michael Marks' first stall in Leeds market
1884	Julius Maggi's powdered soup
1886	Coca-Cola developed, Atlanta, by John Pemberton
1887	Hovis patented and sold as a health food by Richard (Stoney) Smith, Stone, Staffs.
1888	Home and Colonial opened
1889	Board of Agriculture (later MAFF) oversaw agricultural education

1890	Government enabled county councils to provide agricultural education
1890s	Chocolate as confectionery manufactured
1890s	Gas cookers in homes
1892	Smallholdings Act
1894	Wye College became an agricultural college
1894	Coca-Cola bottled, Vicksburg, Missouri
1895	Midland Dairy College established (later to become University of Nottingham School of Agriculture)
1895	Shields's pulsating milking machine
1897	Royal Commission on the state of agriculture
1899	Atwater's calorimetry led to recommended energy requirements
1899	F.G. Garton, Nottingham, developed HP Sauce
1899	Many volunteers for the Boer War found to be undernourished, authorities concerned about food supply
1899	Decker's *Cheese making* first published
c.1900	J. Maggi's beef stock cubes
1901	Harper Adams College opened
1901	British Agricultural Organisation Society formed (first founded in Ireland)
1904	Survey discovered chronic debility and under nourishment
1904	Popularisation of the hamburger in USA
1904	Iced tea invented
1904	National Farmers' Union (NFU) formed in Lincolnshire
1905	Tea bags introduced
1905	*Journal of Agricultural Science (Cambridge)* first published
1906	Pure Food Act to reduce adulteration
1906	Atwater and Bryant's tables of food analysis published
1906	Kellogg's Corn Flakes, Battle Creek, Michigan
1907	Central Land Association formed, later Country Landowners Association
1908	Smallholdings and Allotments Act
1908	Oscar Kellner's starch equivalent animal feeding system

1908	Lyon's Corner House cafes started in London
1908	Invention of the tea bag by Sullivan
1909	Institute of Grocery Distribution (IGD) formed
1911	First recorded hen battery cages, USA
1912	Lord Ernle's *English Farming Past and Present* first published
1912	Vitamins discovered
1913	*Journal of Agricultural Research* (Washington DC) first published
1914-1918	World War I, campaign for food production, county War Agricultural Committees
1914	Milk and Dairies Act prevented sale of milk from tuberculous cows
1915	Discovery of water soluble and fat soluble growth factors (vitamins) for rats
1915	WIs reach Britain, involvement in rural infrastructure
1916	First self service food store, Clarence Saunders, USA
1917	Corn Protection Act to guarantee prices
1917	Haldane Committee created research councils for sciences
1919	Board of Agriculture became Ministry of Agriculture
1919	Land Settlement (Facilities) Act, county council smallholdings set up
1919	Jack Cohen's first market stall (see Tesco 1932)
1919	WI markets started
1920	Agriculture Act provided some protection for the home industry
1920s	Soup kitchens set up to feed the destitute
1920s	Automatic bread slicing machines introduced in large bakeries
1920s	Serious cultivation of sugar beet in UK began
1921	Betty Crocker cake mixes introduced
1921	Protection withdrawn
1924	Rudolph Steiner suggested ideas leading to the biodynamic movement (similar to organic)
1924	Watson and More's *Agriculture* first published
1925	Hen battery cages reached England

1926	National Institute of Poultry Husbandry founded
1927	Glass lined milk tanker railway wagons introduced
1927	Demeter established as a biodynamic trade mark
1927	Last ox team in England sold
1928	Combine harvester first used in England
1928	Harry Ramsden opened fish and chip shop in Guiseley, near Leeds
1928	Heinz baked beans reach Britain from USA, where developed early 19th century
1928	*Journal of Nutrition* first published (USA)
1929	Clarence Birdseye's quick freezing plant
1929	Allied Supplies formed from Lipton, Home & Colonial, Maypole Dairies (see 1819), later to become Safeway (see 1982)
1930	First supermarket, Michael Cullen, USA
1930	Land Drainage Act
1931	Agricultural Marketing Act to set up marketing boards
1931	Agricultural Research Council set up
1932	Tesco formed by Jack Cohen
1932	Boyd Orr's *Food, Health and Income* report, forerunner of the *National Food Survey*
1933	Milk Marketing Board established
1934	Free school milk introduced by the Milk Act
1934	*Farmers' Weekly* first published
1934	Land Settlement Association established to place unemployed on smallholdings
1935	*Food Research* first published
1936	Spam developed by Jay C. Hormel, USA
1937	Restaurant started by McDonald's, Pasadena, California
1938	*Statistical Tables for Biological, Agricultural and Medical Research* first published by Fisher and Yates of Rothamsted Experimental Station
1939-1945	World War II, food shortages in Britain led to rationing
	Plough up policy, supervised by county War Agricultural Executive Committees
1940	McCance and Widdowson's tables of *The Composition of Foods* first published

1940	*National Food Survey* started, originally as the *Wartime Food Survey*
1940	*A Textbook of Dietetics* published by Davidson and Anderson
1940s	Self service stores arrived in Britain
1940s	Scampi brought to Britain by 8th Army servicemen posted in Italy
1940s	Factory canteens and British restaurants increased eating out
1943	Luxmore Committee reviewed provision of advice to farmers
1945	MAFF took over poultry advice from county councils
1945	Brody's *Bioenergetics and Growth* published as part of US drive for agricultural technical efficiency in World War II
1946	Soil Association founded to promote and trade mark organic standards
1946	NAAS set up by MAFF
1946	Ferguson tractor manufacture started in Coventry
1946	Instant mashed potato
1947	Agriculture Act introduced annual price review, established support services
1947	GATT (General Agreement on Tariffs and Trade)
1947	*Food Technology* first published
1949	Sugar rationing ended in post war Britain
1949	Country Landowners Association formed (CLA)
1950	*Journal of the Science of Food and Agriculture* first published
1950s	Poultry hybrids introduced
1950s	First TV cook (Marguerite Patten)
1950s	Delicatessens introduced to Britain
1950s	Demise of the town dairies
1950s	Fish fingers introduced
1951	First National Park designated
1954	End of rationing of poultry mash permitted expansion of production
1954	Broilers introduced to UK from USA
1954	Meat rationing ended in post war Britain
1954	TV dinners introduced in USA
1956	Broiler hybrids introduced to Britain by R. Chalmers Watson and G. Sykes

1957	Agriculture Act modified
1957	Common Agricultural Policy (CAP) introduced in Europe
1958	Chorleywood bread process introduced to speed baking
1959	*Human Nutrition and Dietetics* first published, edited by Davidson and Passmore
1959	First healthy eating cook book published
1960s	Growth of integration and agribusiness, particularly in poultry and horticulture
1960s	Indian and Chinese restaurants grow in popularity
1961	Crown Estate Act established Crown Estate Commissioners
1962	Sam Walton started Wal-Mart, Bentonville, Arkansas
1962	UK Agricultural Students Association founded at Sutton Bonington
1963	*Silent Spring* by Rachel Carson boosted the organic movement
1964	*Animal Machines* by Ruth Harisson popularised animal welfare issues
1965	*Brambell Report* on animal welfare
1965	ASDA formed from Associated Dairies
1967	Foot and mouth disease outbreak
1968	Agriculture (Miscellaneous Provisions) Act, provided for animal welfare etc.
1970s	Growth in popularity of home freezers
1971	NAAS dissolved, ADAS set up
1972	Rare Breeds Survival Trust established
1973	UK joined European Community
1974	World energy crisis stimulated technical efficiency
1980	School milk abolished
1982	Argyll acquired Allied Supplies to form Safeway
1986	Charges for ADAS work introduced
1988	Museum of the British Poultry Industry founded
1989	*Salmonella* in eggs scare
1989	UK Register of Organic Food Standards formed
1990	Food Safety Act introduced with wide powers
1990s	Growth in consumer concern about food

	safety, animal welfare, healthy eating, environmental and planning issues
1990s	Growth in niche markets, e.g. organic, Freedom Foods
1990s	BSE reduced beef sales
1990s	Growth in popularity of ethnic foods
1990s	Greatly increased proportion of food eaten outside the home
1990s	Round the clock food shopping
1990s	Beginnings of on line food shopping
1990s	Influence of TV and celebrity cooks
1990s	Claims that chicken tikka masala overtook fish and chips in popularity
1992	ADAS became commercial agency
1992	MacShary reforms of Common Agricultural Policy begun, aimed at moving prices closer to world levels
1994	General Agreement on Tariffs and Trade (GATT), Marrakesh agreement to liberalise world trade
1994	Milk Marketing Board wound up
1995	Last Potato Marketing Board cropping quota
1997	ADAS privatised
1997	Farmers' markets started operating in Britain
1997	Institute of Grocery Distribution described consumer sovereignty
1998	Four supermarket chains had two-thirds of the grocery trade
1998	Public concern about genetically modified (GM) organisms in agriculture and food
1999	Countryside Commission became Countryside Agency
1999	Food Standards Act published, for establishment of Food Standards Agency, taking over food safety responsibilities from MAFF
1999	EU regulation to ban conventional battery cages in stages by 2012
1999	Announcement of policy switch from production subsidies to environmental
1999	National Association of Farmers' Markets formed

2000	Food Standards Agency (FSA) launched
2000	*The Cambridge World History of Food* published
2000	Countryside Agency *Eat the view* initiative launched
2001	Start of FSA supervision of food standards enforcement by local authorities under 1999 Act
2001	Serious outbreak of foot and mouth disease, with wide economic, social and political consequences
2001	MAFF wound up; Department of Environment, Food and Rural Affairs (DEFRA) established
2002	Curry Report on the future of food, agriculture and the countryside suggested further emphasis on environmental issues

REFERENCES

The pre-historic dates are approximate and vary between sources.

The information in the chronology was gleaned from numerous of the published sources listed in the following chapters, plus:

Barlosius, E. (2000) France. In: *The Cambridge World History of Food*. Edit. Kiple, F.K. and Ornelas, K.C., Cambridge University Press, 1210-1216

Carpenter, K. (2000) Proteins. In: *The Cambridge World History of Food*. Edit. Kiple, F.K. and Ornelas, K.C., Cambridge University Press, 882-888

Dearlove, P. (2000) Fen Drayton, Cambridgeshire: An Estate of the Land Settlement Association. In: *Rural England*. Edit. Thirsk, J., Oxford University Press

Edmunds, J. (1982) *A General History of Uttoxeter*. Uttoxeter & District Civic Society

Hughes, D. and Ray, D. (1999) *Developments in the Global Food Industry*, Wye College, University of London

Ingram, A. (1987) *Dairying Bygones*. Shire Publications, Princes Risborough

Lampkin, N. (1994) *Organic Farming*. Farming Press, Ipswich

Larsen, C.S. (2000) Dietary reconstruction and nutritional assessment of past peoples: the bioanthropological period. In *The Cambridge World History of Food*. Edit. Kiple, F.K. and Ornelas, K.C., Cambridge University Press, 13-34

Miller, N.F. and Wetterstrom, W. (2000) The beginnings of agriculture. In: *The Cambridge World History of Food*. Cambridge University Press, 1123-1136

Miller, W.C. and West, G.P. (1962) *Black's Veterinary Dictionary*. Black, London

Muller, H.G. (1986) *Baking and Bakeries*, Shire Publications, Princes Risborough

Nestle, M. (2000) The mediterranean. In: *The Cambridge World History of Food*. Edit. Kiple, F.K. and Ornelas, K.C., Cambridge University Press, 1193-1203

Pellett, P.L. (2000) Energy and protein metabolism. In: *The Cambridge World History of Food*. Edit. Kiple, F.K. and Ornelas, K.C., Cambridge University Press, 888-913

Roe, D.A. (2000) Vitamin B complex. In: *The Cambridge World History of Food*. Edit. Kiple, F.K. and Ornelas, K.C., Cambridge University Press, 750-754

Twiss, S. (1999) *Apples: A Social History*. The National Trust

Wolf, G. (2000) Vitamin A. In: *The Cambridge World History of Food*. Edit. Kiple, F.K. and Ornelas, K.C., Cambridge University Press, 741-750

Additional sources were:

Product labels, advertisements, brand publicity material and numerous web sites.

PREFACE

In normal times, in an industrial and post industrial economy, most people are not much interested in what goes on, still less what used to go on, in agriculture or in its supporting sciences. At the time of writing, however, things are a little different and these are not normal times. The place of agriculture, food and the countryside in modern Britain is a regular talking point. National strategic reviews of the subject are under discussion at the time of writing, so some background information on the story of the evolution of our food, and of our agricultural and countryside heritage, may be of interest. It is not the first time in history that rural strategy has been a national issue.

Everyday life owes more to the history of food and farming than readily meets the eye. I go so far as to agree with those who have claimed that the advent of civilisation itself depended on two significant events in agricultural history: the domestication of farm animals and that of the cereals. Such claims are not new: they were mentioned in undergraduate agricultural botany courses of the 1950s, but they have been forgotten in modern excitements about more glamorous inventions. More recently some have gone further and suggested that the history of food determined the history of the world, though others have argued that the influence was not always wholly for the better.

The first chapter deals with the period of recorded history, but recorded history is not the whole story. Therefore subsequent chapters deal with some key events which took place much earlier and made history possible. As an aside the history of pigs and poultry is used as a sector example in slightly more detail. This example is chosen partly because it illustrates the response of an agricultural sector to changing times, and to changing demands made on it by consumers and the public, but also because it is a branch of agricultural history I happen to know a bit about.

Little original material has been researched, since that is not the main point of the book. However a large amount of material has been compiled, summarised and discussed, and references to sources are numbered throughout the text, with a list at the end of each chapter. This may, perhaps, seem a little ponderous, but at a time when food, farming and the countryside are sometimes discussed in terms of whims and fashions some quoted sources may be useful. Anyway, I cannot escape the habits of 40 years of scientific work.

David Charles
2002

1

AGRICULTURE, FOOD, AND EVERYDAY LIFE IN BRITAIN

The general great advantages to the public of improvements in agriculture, and the oeconomical arts, are too evident to require any demonstration from argument. But there are reasons for our earnest attention to them, peculiar to the present times, which cannot be too strongly and universally inculcated.

(R. Dossie, 1768)[1]

THE HISTORY OF THE ORDINARY

Much of history is about kings, queens and political events, and rightly so because such things created the structure of the country in which we live. But for most of the time, except during invasions and political upheavals, the vast majority of the ordinary people probably scarcely noticed these things. Now and again the ordinary citizen was dragooned into taking part in some military campaign, territorial reorganisation or religious enthusiasm which he only dimly understood. But the important issues in the rural areas were the state of the fields, the prospects for the harvest, and the competence of the village swineherd.

As Julian Wiseman pointed out in his *History of the British Pig,* even agricultural histories tend to be incomplete, having little to say about the pig on which so many of the peasantry depended for their meat.[2] He added that this was because the illiterate classes were not much given to writing agricultural textbooks. I would add that, for similar reasons, there are three more omissions from many of the textbooks, particularly those written before about 1850.

Firstly poultry do not receive much prominence, yet in medieval London eggs and poultry were on sale.[3] It is interesting that some of the detail about this trade turns up in a history of shopping, not a history of agriculture or of food. Even if not prominent in agricultural texts, poultry dishes feature in accounts of 18th century cookery. A 19-course meal, nine of which were poultry, from Mrs Frazer's *The Practice of Cooking,* was quoted in a book on Jane Austen.[4]

Secondly agricultural histories sometimes have little to say about fruit and vegetables. This is because despite the present day enthusiasm of dietary writers for them, in former times they were regarded as only good enough for labourers and servants. In medieval times some even suspected that raw fruit might be dangerous. Was this the first British example of a food scare? Although both cottages and great houses had kitchen gardens the word vegetables was not even used. *Pot herbs* was the common term, and the better off just called them *garden stuff*.[4] Root vegetables, since they were regarded as earthy, were relegated to the diet of the peasantry, and therefore they were not fashionable with early food historians.[5]

Thirdly we hear little in the agricultural texts of the past about the lives of the landless in the English countryside. We learn more about them from the likes of Thomas Hardy in fictional work; from William Cobbett, and from Joseph Arch's autobiography. If we are to understand properly the countryside and its heritage such omissions may be important in view of today's occasionally rather romantic and nostalgic images of its past.[105, 107]

We should not be particularly surprised if the emphasis on agricultural matters in the recording of national history gets less and less rather than more and more. As modern life gets further from the land awareness of these things gets understandably less. In former times most people, even in the towns, knew and recognised the individual farmers who sold them their food, and could see the fields from whence it came. It is harder to associate a packet on a shelf with a farm of origin, though modern practices, including traceability to source, farm shops and farmers' markets are turning this full circle. Branding is a modern method of letting consumers identify producers when personal contact is impractical.

It might be interesting, or even useful, to attempt occasionally to take a few different perspectives on our food and rural heritage. Such perspectives can raise potentially iconoclastic questions. For example a modern urban Briton, asked for his nomination for the most consequential human invention of all time would, I suspect, probably vote for something like the internal combustion engine, the silicon chip, the mobile phone, or television. But I agree with those who have contended that the answer is the domestication of farm animals and of cereals, without which none of the rest would have been possible. Admittedly the domestication of wheat would not have got civilisation very far without a few accessory inventions such as the plough and bread making, as we shall see later.

The declining importance of agriculture in the national life is real and is illustrated by a few figures on the cost of food in the family budget. The Ministry of Agriculture, Fisheries and Food (MAFF), and now the Department for Environment, Food and Rural Affairs (DEFRA), publish statistics on the percentage of consumers' total expenditure spent on household food. Their

1994 *National Food Survey* quoted Central Statistical Office figures suggesting that in the 1950s it was about 28%.[6] By 1989 it had fallen to 13% and by 1999 to 10%.[7] However household food accounts for only a part of today's total food spending because eating outside the home is now significant. Thus total food spending as a percentage of consumers' expenditure has been estimated at 26% in 1970, falling to 17% by 1999/2000.[8] There are, of course, large ranges around these averages.

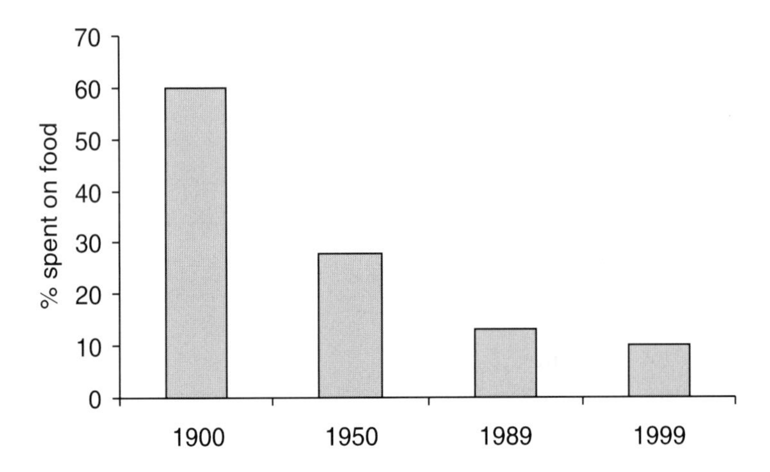

Figure 1.1 Percentage of consumers' total expenditure spent on household food

At the beginning of the 20th century keeping a family fed must have been a much bigger part of the family budget and a very important matter for many people. In 1901 the average worker spent an astonishing 60% of his income on food.[9] In the early 19th century food seems to have been even more expensive. A side of bacon was worth including in the auction at a house sale in 1801.[4] Modern food is not only cheap: it is also convenient. Much of the price of modern food is value added to it in the form of processing and packing after it has left the farm. Modern consumers are not content with raw materials.

Another statistic illustrating the declining prominence of the land in the economic lives of most of us in Britain is the proportion of the population engaged in agriculture. According to some figures summarised by Sir Kenneth Blaxter and Noel Robertson, in the period from 1695 to 1700 about 24% of the population was agricultural. Add to these their dependents and it was probably at least half of the British population who lived close to the land and depended on it for their survival. By 1866 the figure had fallen to 6%, by 1936 to 2% and by 1986 to 1%.[10] MAFF figures for 1998 noted that only 615,000 people out of over 58 million of us in UK were employed in agriculture.[11]

By contrast agriculture employs about 46% of the economically active population world wide.[12]

While it is true that only a minority of modern Britons depend on the land for their living, many more wish to enjoy the countryside as an amenity.

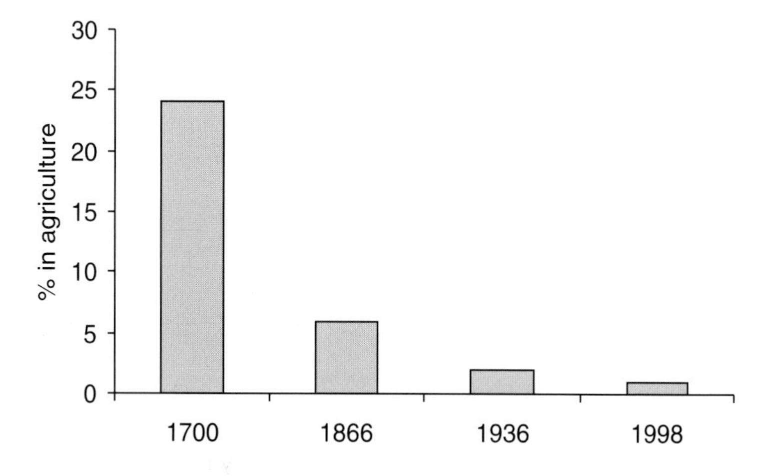

Figure 1.2 Proportion of the British population engaged in agriculture, %

Some wish to live in it, so it is still an issue in modern Britain. Its history is therefore a background to some modern concerns.

Agriculture is part of the food supply chain. Until relatively recently it constituted most of the food chain, but in modern times more and more of the final money value of our food is added after it leaves the farm. Thus, despite the small numbers now involved in primary agricultural production, that is now far from the end of the story. The British food and grocery industries employ several million people.

Because agriculture is part of the food chain it is impossible, and inappropriate, to separate agricultural history from the history of food, and histories of food contain much overlap between the two.[13,14] Since some of the medieval explorations were in search of spices, and since trade in food and the need to supply armies has dominated many important historical events, to some extent the history of food could be said to have influenced the history of the world. Chocolate, tea, coffee and spices dominated international trade for centuries. Local and national food supplies were so vital (literally) until recently that wars were usually delayed until after the harvest.[14] The supply of wheat to Rome shaped a good deal of the politics and organisation of the Roman empire, and thus the history of European civilisation.

RURAL HISTORY - THE ERA OF GREAT FARMERS AND CHRONICLERS

Some agricultural writers and journalists

What were all those people doing in the days when so many were on the land? A number of writers have chronicled the history of agriculture, even if some of us believe that they might have undervalued pigs, poultry, fruit and vegetables. Arthur Young (1741-1820) is usually credited with having been the first significant British agricultural journalist, though agricultural textbooks originated in much earlier times. Young did not consider himself the first, and in the Introduction to Volume I of his *Annals of Agriculture* in 1784, he referred to Houghton's regular agricultural paper of a century earlier.[15]

Much earlier still a *Farmer's Almanac* was produced by Sumerian scribes by about 2500 BC according to Tannahill.[16] This must have been about the earliest agricultural literature of all, though descriptions of farming matters also frequently feature in Egyptian art and writings. The Greek writer Hesiod offered advice to small farmers in the 8th century BC. His advice had a firmly practical tone to it, despite the fact that he is primarily remembered as a poet. In his *Works and Days* he recommended keeping a spare plough in case one got broken, which was probably because he was much concerned with performing field cultivations on time. He recommended that oxen for ploughing should be nine years old and that a ploughman should be 40 years old. At 40 the ploughman was considered young enough to work, but no longer subject to the distractions of youth. He also recommended choosing a woman for her ability to plough rather than as a wife. Not surprisingly his works have been translated and commented upon many times.[17, 18]

But probably the earliest specialist and comprehensive text of agricultural advice to come down to us in full is that of the 1st century AD Roman writer Lucius Junius Moderatus Columella, translated from Latin into English in 1745. His book is really a collection of 13 books covering everything connected with cropping and animal husbandry, right through to the pickling and preserving of produce.[19] So comprehensive were his books that they even included a chapter quaintly entitled *"With what Remedies the Pain of the Belly, and of the Intestines of Cattle, may be quieted".*

The translator of Columella considered the subjects important enough to make the following extravagant claims in the preface:

The Art of Husbandry is so necessary for the support of human life, and the comfortable subsistence and happiness of mankind have so great a dependence upon it, that the wisest Men in all ages have ascribed its original to God, as the Inventor and Ordainer of it; and the wisest and most civilised Nations, who have understood their true interest, have always endeavoured to promote and improve it; and have never failed to acknowledge, and honour, as public Benefactors, all such as contributed any thing towards the same".

We shall examine later how such attitudes have evolved since these sentiments were recorded.

Columella is the best known Roman agricultural author but he was not the only one. Palladius wrote 14 books on husbandry in the 4th century, and these were translated into English by the Rev. Barton Lodge in 1873, from manuscripts of about 1420 found in Colchester Castle. The 1420 version appears to have been in verse. Palladius recognised the value of diligent management, so that Lodge's 1873 edition states that "The master's presence benefits his fields". The original 1420 version had apparently put it more archaically, if more picturesquely, as "The lord present his feelde may best avaunce".[20]

An early example of an English agricultural textbook with ambitious objectives was that of Matthew Peters published in 1771: *The Rational Farmer: or a Treatise on Agriculture and Tillage*. The cover page proclaimed: "Wherein many errors of common management are pointed out, and a new improved and profitable System suggested and described; interspersed with many occasional and interesting observations".[21]

In 18th century England the study of agricultural matters was considered a worthwhile occupation for those in a position to do it. In 1762 J. Mills wrote of his opinion that: "Philosophical inquiries into the principles of vegetation, are an object well worth attention of gentlemen whose situation allows them to pursue that truly useful and entertaining study".[22]

Thus there were many farming textbooks published in the 18th and 19th centuries: a convenient fact for modern agricultural historians. Some of the books were repeatedly reissued as subsequent editions, so that their names survived over long periods. An example of one which survives into modern times is Primrose McConnell's *Agricultural Notebook*, originally published in 1883 and now in its 19th edition.[23] Another is *Fream's Elements of Agriculture*, first published in 1892 and now in its 17th edition, renamed *Fream's Principles of Food and Agriculture*, reflecting the modern emphasis on food.[24] An early 20th century example textbook was Watson and More's *Agriculture, the Science and Practice of British Farming,* first published in 1924.[25] Some 19th century works survived less well as textbooks, but their illustrations are worth preserving as art. Examples include Andrews's *Modern Husbandry* of 1853[109] and Youatt's *Cattle* of 1834.[110]

William Cobbett, whose *Rural Rides* was published in 1830, was one of the best known chroniclers of the early 19th century English countryside. As a

Illustrations in text books as art: an example from Andrews's *Modern Husbandry*, 1853[109]

farmer's son turned writer he toured the country during the depressed 1820s, often defending those who faired worst in those troubled times, and once imprisoned for sedition. More of him and his writings later.[107]

An important agricultural history, used as a source by many later commentators, was *English Farming Past and Present,* by Rowland Prothero (Lord Ernle), first published in 1912 and revised several times by Prothero himself until 1936, and once more by other authors in 1961.[26]

The Royal Agricultural Society of England, founded in 1838 and granted a royal charter in 1840, has published an annual journal ever since that time. Key food and agricultural histories written in modern times include Tannahill's *Food in History*, first published in 1973 and revised in 1988.[16] Toussaint-Samat produced a large volume in French in 1987 on the history of food, and it was translated into English by Bell in 1992 and re-issued in paperback in1994.[14] Sir Kenneth Blaxter and Noel Robertson wrote an account in 1995 of agricultural history from the point of view of technical progress and thus of the consequent benefits to consumers.[10]

Readers did not always appreciate the opinions of the writers and experts. It seems that in Ptolemaic Egypt there was an occasion when the practical peasant farmers on an estate of 10,000 arourae (about 82,000 hectares or 200,000 acres[27]) complained that blunders had occurred because more notice had been taken of Greek experts than of them.[28]

Changing times

In Roman times agricultural settlements in Britain were based on the villas of the ruling families, but from Saxon times onwards English villages were organised on the communal open field system, where one field out of three was left fallow every year in order to recover some of its fertility. The three field system was occasionally replaced by a two field system. Within the fields the villagers held random strips of land in all three fields, rather than having individual farms. In many parts of the country, particularly in the Midlands, the layout of the little strips of cultivation can still be seen in the fields, particularly during dry summers, as ridges and furrows caused by hundreds of years of ploughing in the same direction. An acre was originally a day's work.[113]

The ridges and furrows can still be seen in this field at Willoughby on the Wolds, Nottinghamshire, where strips of land were cultivated throughout the Medieval period. From 1743 to 1869 Parliament passed 154 Enclosure Acts for Nottinghamshire (Lyth, 1989).[111] These fields at Willoughby were enclosed in 1793 (Atkins *et al.*, 1999).[32]

Disadvantages of the open field system included rather rigid ideas and rules on rotations, sometimes unsuited to the local soils. The separation of the arable and pasture land missed the benefits of mixed rotations. The more successful and the more influential participants, and the manorial landlords, became ambitious to have more land and independence, and to have their land more conveniently located around their homesteads. The result was enclosure, which gradually created larger fields. Enclosure reached different parts of the country at different times, but was probably complete by about 1815. The parish of Laxton in Nottinghamshire survives as the last remaining entire village example of the open field system.[29,30] Strip cultivation has survived elsewhere, however. In Epworth, Belton and Haxey (Isle of Axholme, Humberside) some of the land is still farmed this way because the small holders of the 18th and 19th centuries were unwilling to pay the costs of enclosure.[31] A detailed study of the processes and the consequences of enclosure in a south Nottinghamshire village, by Atkins, Hammond and Roper, serves as an example of events going on throughout England in the 18th and 19th centuries. They noted the roles of the commissioners in allocating land between claimants, and of the surveyors in practical measurement.[32]

The open field system at Laxton, Nottinghamshire.

The fallow year of the three or two field system was to permit the soil to regain some of its fertility, though the importance of returning manures to the land had long been understood. Columella wrote in his *Book the Second* "That the Earth neither grows old, nor wears out, if it is dunged". Palladius valued poultry for their manure "whose dung is necessary for the land, except that of the goose…"[20] The Greeks valued fallow land, and knew about the benefits of alternating cereals and legumes. Hesiod regarded fallowing as a kindness to the children.[17,18] The ancient Egyptians did not need to worry about fallowing, rotations or manuring, however. The flooding of the Nile deposited so much rich silt that three crops a year were often possible.[28]

Radical technical developments in English agriculture included the introduction of turnips and clover by the 16th and 17th centuries, and alternative husbandry with a grass interval in Tudor times. Cropping and the care of the land reached a famous high point with the Norfolk four course rotation (wheat-turnips-barley-clover). The development of it is usually associated with famous pioneers like Turnip Townshend (1674-1738) and Coke of Holkham (1754-1842);

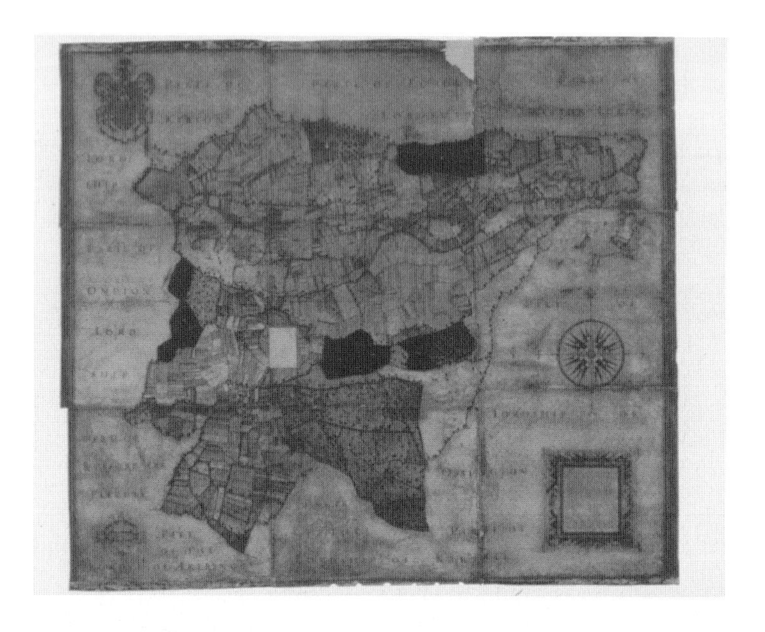

Pierce's map of Laxton in 1635. Reproduced with permission of the Bodlein Library, University of Oxford.

Detail from maps of Laxton in 1635. Reproduced with permission of the Bodlein Library, University of Oxford.

A map of Laxton after the limited enclosures of 1908. Reproduced with permission of the Trustees of the Laxton Visitor Centre.

though Blaxter and Robertson made the valid point that few developments really depended on one charismatic individual, but on gradual evolution.[33,10] The famous figures were probably popularisers rather than pioneers, but that is a worthy enough achievement. A certain Mr. Brett of Norfolk deserves a mention for his unintentional contribution to rural history. He was a tenant on the Holkham estate who refused to pay 5 shillings per acre rent. Thus the young Thomas Coke decided to run the farm himself, and became fascinated by agricultural problems.[34]

The turnip: a crop which contributed to the revolution in rotations.

Hand driven root crop cutters like this were still in occasional use until the 1960s. This one was probably made in the 1940s or 1950s.
Photo taken at the Farm and Country Centre, Sacrewell, Peterborough, with permission.

Another of the famous figures of that era was Jethro Tull (1674-1740), associated with the horse hoe and the seed drill and their use in higher standards of husbandry. His machines permitted sowing in rows at the correct depth, followed by hoeing to control weeds, instead of tolerating unweeded crops sown at random.[35] He thought that a benefit of field operations like hoeing was the creation of tiny soil particles to be taken up by crops, but despite this misconception his achievements were considerable. Some of his contemporaries, such as Duhamel, agreed with his notion that "very minute particles" were the food of plants. Duhamel wrote "*A Practical Treatise on Agriculture*" in 1762.[36] Jethro Tull may not have realised that weed control was one of his best achievements, but he did thoroughly understand its importance, and was forthright in his abomination of weeds. How about the following for strong words:

> *Plants that come up in any Land, of a different Kind from the sown or planted Crop, are Weeds.*
> *That there are in Nature any such things as inutiles Herbae, the Botanists deny; and justly too, according to their Meaning.*
> *But the Farmer, who expects to make a Profit of his Land from what he sows or plants in it, finds not only Herbae inutiles, but also noxiae, unprofitable and hurtful Weeds.*

Descendants of Bakewell's Longhorns at a tableau at Leicestershire Show, 1995.

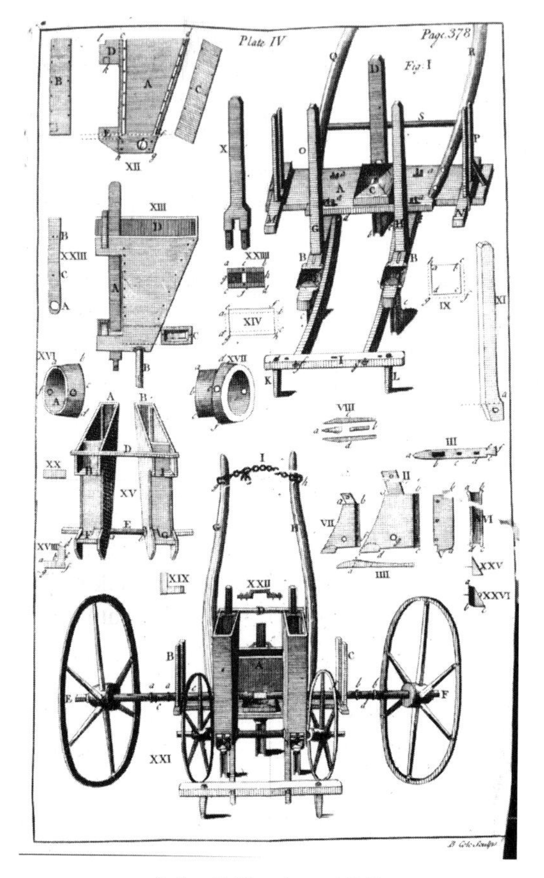

Jethro Tull's wheat drill.[35]

Later came the great livestock breeders like Robert Bakewell (1725-1795) and the Colling brothers.[37,38,39] They have achieved recognition of a sort in everyday modern England. To this day many a pub is called *The Durham Ox* after the Colling brothers' famous beast.

Less famous people also made significant contributions. For example as early as 1594 Sir Richard Weston had published observations he had made in the Netherlands on the use of root crops and clover.[26] Sir Henry Hobart of Blickling and Felbrigg, Norfolk, probably also deserves a mention amongst the greats. His improvements as early as between 1596 and 1625 concentrated on estate organisation.[40]

Many commentators have claimed, probably justifiably, that these great agriculturalists made the industrial revolution possible by making it feasible to feed the rapidly expanding urban population. From the end of the 17th century to the mid 19th century the national population increased from 6 million to 21 million, so feeding them mattered.[10]

The 19th century saw some great surges of progress in the mechanisation of field operations, taking Jethro Tull's work much further than he could have imagined. The harvesting of the cereals had been a manual operation, using sickles, since the domestication of cereals in about 9000 BC. Until the 19th century the annual harvest was the work of small armies of labourers. Mechanised reaping had been attempted in the Netherlands much earlier, but not in Britain. In 1577 Barnaby Googe wrote *Foure Bookes of Husbandry*, quoted by Prothero in his agricultural history of 1936.[26] Googe's writings included picturesque accounts of the astonishing machines he had seen in the Netherlands. He appears to have had difficulty believing his own eyes when shown a mechanical reaper cutting corn.

In 1828 Patrick Bell began a process of development which was to change harvesting forever, with enormous consequences for both efficiency and social effects. He devised the first British reaper, though there had been earlier patents by James Smith.[41] Cyrus McCormick's reaper of 1831 was more commercially successful in the new world, eventually reaching England in 1851. Writing in the *Journal of the Royal Agricultural Society* for that year Pusey declared that it "...cut the wheat...with the utmost regularity."[42] By 1871 McCormick's Chicago based company was turning out 10,000 harvesters a year. The first combine harvester used in England was in 1928. The Royal Agricultural Society of England was active in encouraging engineering developments, and at their first show at Oxford in 1839 Robert Ransome, with his self sharpening plough share and many other items of field equipment, was awarded a gold medal.

Steam power began to replace animal and human muscle power in the 19th century. The first patent for a steam plough was by Major Pratt in 1810, eventually leading to Fowler's double engine ploughing tackle in 1860.[41]

Agricultural engineers were also influential during the 20th century. In 1906 there were a million working horses on the farms of England and Wales. By 1939 there were only half as many, but there were 55,000 tractors. There are accounts of ox teams still at work into the 1920s in Sussex. By the 1960s tractor numbers had increased to 390,000.[43,44] Tractor numbers remained fairly similar through the 1970s and beyond, but they became much more powerful in order to achieve quick and timely tillage operations, thus contributing to the efficiency of UK agriculture as measured in fundamental physical and biological terms, such as the capturing of solar energy by crop plants.

The Women's Land Army on field work in 1939. Photo G.S. McCann, Uttoxeter. From the collections of the Department of Manuscripts and Special Collections, University of Nottingham.

It was machines like this reaper and binder, with its string knotting device,which helped to revolutionise harvesting.

Steam engines like this one hauled massive ploughs at the height of the steam age, but setting up for the day's work must have been a tiresome business.

A demonstration of steam powered field operations.

No account of the contribution of engineering would be complete without a mention of the little grey Ferguson TE20 tractor with its three point implement linkage and hydraulic lift. Once a routine part of the farming scene it is now popular with collectors and enthusiasts. Some are still in use. Manufacture of the TE20 by Harry Ferguson started at Coventry in 1946.[45]

A line up of Ferguson tractors at an enthusiasts' event in 2002.

SOCIAL BACKGROUND

For 200 years there have been political debates and swings in policy on the question of government intervention in agricultural prices, and the protection of the home industry from cheap imports. Part of the debate concerns the question of whether or not farms are like any other businesses, and thus whether or not it matters if the successful become larger, and eventually corporate, while the unsuccessful go to the wall. As long ago as 1801 William Wordsworth wrote to the Whig leader Charles James Fox about, among other things, the plight of the disappearing lakeland freehold farmers. Wordsworth held the view that a little tract of land encouraged domesticity.[46] A few years later Cobbett wanted to see the re-establishment of the cottager class as a foundation of rural society.[107] Modern concerns include debates about the effects on the appearance of the countryside and on wildlife.

Over recent years the number of agricultural holdings has been falling, as Table 1.1 shows, though the fall has not been as fast as in the poultry sector, whose history is used as a detailed example later in the book.

TABLE 1.1 NUMBERS OF AGRICULTURAL HOLDINGS, 000s

1955	*1967*	*1984*	*1988*	*1993*	*1998*
376.3[1]	306.6[1]	252.1[2]	247.3[3]	244.2[4]	237.9[4]

[1] Marks and Britton (1989) (holdings of 1acre or more, except where agricultural activity small)[47]
[2] MAFF (1989)[48]
[3] MAFF (1994)[49]
[4] MAFF (1999)[11]
(MAFF data refers to holdings of 1 ha or more of farmed land, or any holding with significant agricultural output)

It has frequently been argued that free trade in food commodities leads to hardship and depopulation in the countryside and unreliable food supplies for consumers. This point of view is founded on ancient ideas. The medieval civic authorities in English towns, including London, were so concerned about possible food shortages and high food prices that they vigourously supervised and controlled trading in food. Bread prices were often fixed and middle men were viewed with deep suspicion, even when some town populations became far too big to be supplied by farmers selling direct.[3] Wool could only be sold in staple towns which controlled the market. There is some evidence of state intervention in agricultural prices even earlier. The Roman emperor Diocletian fixed the prices of British produce in AD 301.[50] It is probable that what he wanted was to budget the cost of feeding Rome and its armies on the march, rather than to assist British farmers with their business planning.

The concept of international free trade is relatively new, dating from Adam Smith's book on *The Wealth of Nations* in 1776.[65] Since Adam Smith it has regularly been argued that food, like any other product, should find its market price, and consumers in an industrial country should be entitled to buy it wherever it is cheapest in the world. Since that time the view has often been expressed that farms should be regarded as businesses like any other. For example in 1898, after claims that the railways were exploiting their bargaining strength too much in the carriage of produce, Lord Emlyn, the chairman of the Great Western Railway Co., said that farmers should organise themselves and learn business as well as farming.[51] Practical texts on the farm as a business came much later, and a famous example was *Farming for Profits* by Dexter and Barber, published in1961.[52]

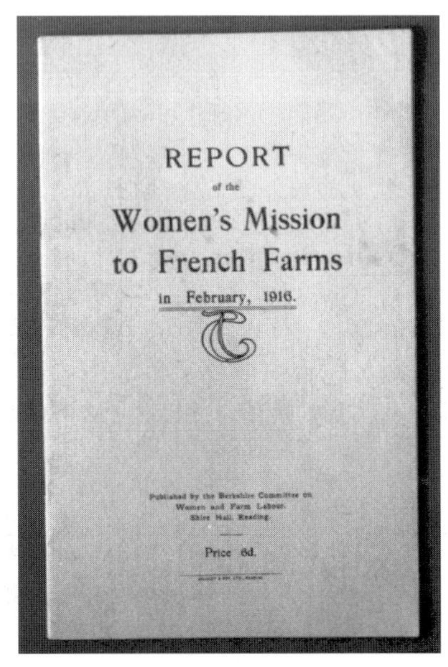

REPORT

of the

Women's Mission
to French Farms

in February, 1916.

Published by the Berkshire Committee on
Women and Farm Labour.
Shire Hall, Reading.

Price 6d.

Such was the importance of agriculture in World War I that a delegation was sent to France to see how French country women were coping with farming after their men folk had gone to war.

Wars and their aftermaths seem to have had important effects on the balance of this debate. Of the famous Corn Laws, that of 1815 was the most important, providing protection for the home industry during recovery from the Napoleonic wars. The protection was not enough, however, to prevent the catastrophic fall of wheat prices in the years after 1815.[13] Numerous amendments, in particular in 1828, and then the repeal of the Corn Laws in 1835 and 1846, removed protection from imports of cheap cereals, mainly from the New World, and agricultural depression followed.[10,13,105] Some have claimed that immediately after 1846 there was not as much effect as might have been expected because demand increased, due to factors such as the growth in the urban population. Yet even by 1844 the Duke of Richmond was leading protests by landlords and tenant farmers against the activities of the Anti-Corn Laws League. Protection Societies were formed.[53] It is generally agreed that after 1874, and free trade, serious rural depression set in, reaching its deepest by the 1880s, as overseas produce began to enter in volume.[54]

The effects of the Corn Laws on the landless rural poor were complicated. On the one hand the protection of home cereal production from foreign competition was presumably intended to protect the commercial viability of farming and therefore of rural employment. On the other hand high bread prices under protection, particularly in bad harvest years, led to hardship, often to the brink of starvation. By 1833 the state of affairs in the countryside

The Women's Land Army in 1939. Photos: GS McCann, Uttoxeter. From the collections of the Department of Manuscripts and Special Collections, University of Nottingham.

More scenes of the activities of the Women's Land Army during World War II. From the collections of the Department of Manuscripts and Special Collections, University of Nottingham.

was already bad enough to bring the field labourers to desperate straits. High bread prices meant poverty and dependence on charity for many.[105] Their early attempts to improve their lot culminated in the transportation of the Tolpuddle Martyrs from Dorset to Australia for a technical contravention of the laws on assembly. Joseph Arch experienced at first hand as a child the desparation of the field labourers, including his father, in Warwickshire in the 1830s and 1840s. His father had refused to sign a petition in favour of the Corn Laws and so jeopardised his employment prospects. Years later these experiences led Joseph Arch to found the first union for agricultural workers in 1872. Some consider that the condition of the field labourers was slightly better in counties where hiring was by the year, than in the counties where hiring was on a week by week basis.[106] Despite improvements after Joseph Arch's activities the countryside remained a place of low wages, and by 1900 agricultural labourers were still the poorest paid wage earners.[13]

Commenting at the time the Duke of Richmond did not blame the farmers for the plight of their labourers in the early 19th century, since he claimed that the farmers were themselves "nearly crushed by the difficulties of the times".[53] But the fate of the agricultural poor after 1815 was complicated by the Poor Law arrangements. Gilbert's Act of 1782 had given parishes the responsibility to provide work for the unemployed, and in 1795 the notorious Speenhamland system, named after the Berkshire village whose magistrates devised it, provided relief in the form of allowances. These were to compensate for low wages and were based on the price of bread and the number of children. Not surprisingly this had the effect of keeping wages down, yet, as Cobbett recorded with irritation, farms were sometimes being neglected.[13,106] He was annoyed to see road gangs doing what he regarded as unnecessary tasks instead of farm work. In 1834 the Poor Law reforms replaced allowances with workhouses and relieved parishes of the responsibilty for finding work.[13]

In some parts of the country small farmers sometimes took on supplementary sources of income like inn keeping. There was emigration of spare younger sons to the New World where they probably hastened the development of its agriculture. For much of the time during the depressed years livestock enterprises were not as badly affected as arable cropping, and in many parts of the country there was therefore a swing to dairying. A perishable product like milk was less vulnerable to competition from cheap imports, and the railways permitted supplies to be sold in London from all over the country. Initially churns were loaded onto trains from local stations, but by 1930 bulk tanker railway wagons were carrying milk to London from dairy companies organising local collection from farms.[55]

There was a slow agricultural recovery up to World War I and then in 1917 the Corn Protection Act provided price guarantees. This was repealed in 1921 and there followed another depression lasting into the 1930s when marketing boards for some of the main commodities began to be set up under the Agricultural Marketing Act of 1931. Some of these marketing arrangements lasted into modern times. The Milk Marketing Board for example was not wound up until 1994 and the Potato Marketing Board in 1995. In 1965 the historian A.J.P. Taylor commented that these support policies were intended to help to maintain rural communities, yet agricultural employment still declined.[56]

Apart from marketing boards and subsidies another device frequently, but less successfully, attempted in the hope of alleviating depression and slowing the consequent drift of population from the rural areas, was the encouragement of smallholdings. The Small Holdings Act of 1892, the Smallholdings and Allotments Act 1908 and the Land Settlement (Facilities) Act 1919 all empowered county councils to purchase land to let to smallholders. Some 31,000 tenants were accommodated but not usually very profitably.[54] The Land Settlement Association of 1934 also established smallholdings.[112] Some of the schemes lasted into modern times, and most towns still have allotments. Effectively allotments go back much further than 1892, however, since at the time of enclosure those dispossessed of land were often permitted to use corners and wastes in fields for hand cultivation. They often grew potatoes on these little plots, thereby achieving some independence of food supply despite having become dependent on wages for the first time.[57] As early as 1795 the Board of Agriculture recognised the importance of this and offered potato husbandry advice for application on such allotments.

Not all allotments are in urban areas. This one adjoins fields at Burnham Market, Norfolk.

The sorry state of the land and of food supply, as affected by both the economy and the condition of agriculture, was the subject of a Royal Commission in

1897.[58] The state of food supply again impressed itself on the authorities when recruiting started for World War I. Only one in three conscripts was found to be fit to fight and undernourishment accounted for much of this poor health.[59] Central government began to take a new interest in both nutrition and agriculture. For example school children were given free milk from 1934 to 1980. Cod liver oil was issued to children from 1939 through the years of World War II.

By the outbreak of World War II the UK had become very dependent on imported food, so that when shipping convoys were at a premium the public suffered real food shortage and rationing. The government introduced a plough up policy which was estimated to have brought nearly 4 million acres (about 1.5 million hectares) into arable production between 1939 and 1941, thus saving 22 million tons of shipping by the end of 1941.[56] The county War Agricultural Executive Committees carried out supervisory and resource assessment functions.[26] The Ministry of Food distributed advisory material on food and offered tips on eking out supplies with specially designed recipes. They broadcast hints and recipes on the radio (then called the wireless). *Dig for Victory* became a famous campaign encouraging vegetable growing in gardens and on allotments. Country dwellers were encouraged to keep pigs.

It was not only official bodies who acted to support and preserve wartime food supplies. The Women's Institute published recipes designed to spin out scarce ingredients, in particular treasures like butter. They advised on the cooking and making acceptable of unfamiliar emergency commodities like whale meat. They made prodigious quantities of jam and preserved fruit, sometimes at special centres, and published advertisements for bee keeping equipment.[108]

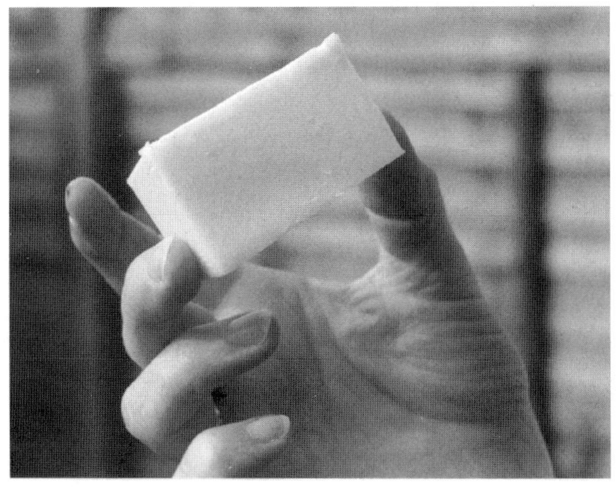

The cheese ration for an adult for a week in World War II.

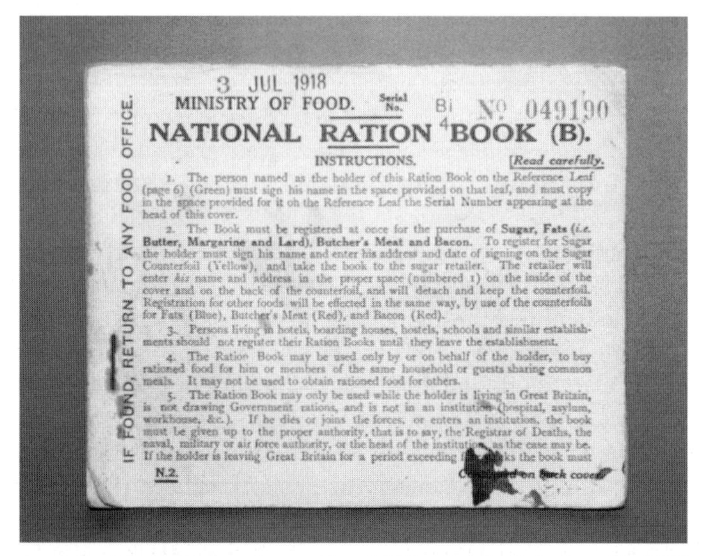

3 JUL 1918

MINISTRY OF FOOD. Serial No. Bi 4 No 049190

NATIONAL RATION BOOK (B).

INSTRUCTIONS. [*Read carefully.*

1. The person named as the holder of this Ration Book on the Reference Leaf (page 6) (Green) must sign his name in the space provided on that leaf, and must copy in the space provided for it on the Reference Leaf the Serial Number appearing at the head of this cover.

2. The Book must be registered at once for the purchase of **Sugar, Fats (*i.e.* Butter, Margarine and Lard), Butcher's Meat and Bacon.** To register for Sugar the holder must sign his name and enter his address and date of signing on the Sugar Counterfoil (Yellow), and take the book to the sugar retailer. The retailer will enter *his* name and address in the proper space (numbered 1) on the inside of the cover and on the back of the counterfoil, and will detach and keep the counterfoil. Registration for other foods will be effected in the same way, by use of the counterfoils for Fats (Blue), Butcher's Meat (Red), and Bacon (Red).

3. Persons living in hotels, boarding houses, hostels, schools and similar establishments should not register their Ration Books until they leave the establishment.

4. The Ration Book may be used only by or on behalf of the holder, to buy rationed food for him or members of the same household or guests sharing common meals. It may not be used to obtain rationed food for others.

5. The Ration Book may only be used while the holder is living in Great Britain, is not drawing Government rations, and is not in an institution (hospital, asylum, workhouse, &c.). If he dies or joins the forces, or enters an institution, the book must be given up to the proper authority, that is to say, the Registrar of Deaths, the naval, military or air force authority, or the head of the institution, as the case may be. If the holder is leaving Great Britain for a period exceeding [illegible] weeks the book must

N.2. [illegible] on back cover

IF FOUND, RETURN TO ANY FOOD OFFICE.

In times of war, food has sometimes been rationed. This ration book is from World War I.

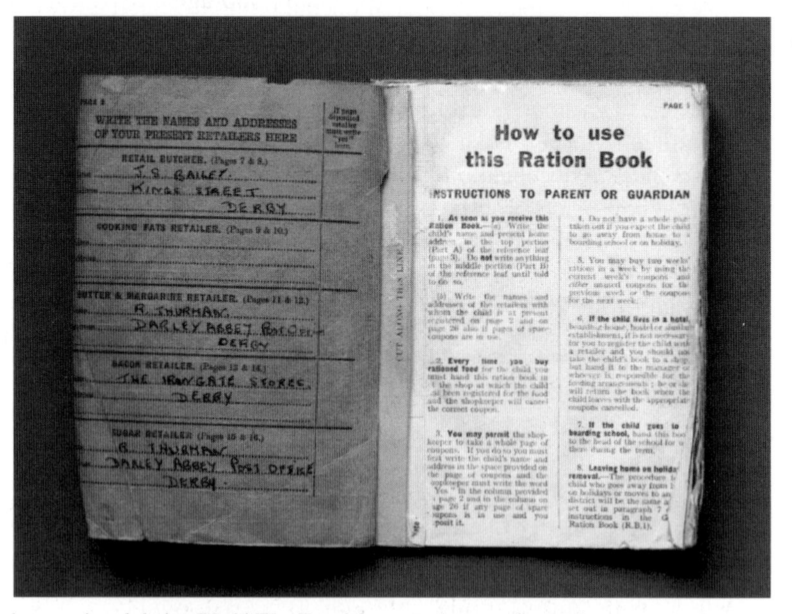

WRITE THE NAMES AND ADDRESSES OF YOUR PRESENT RETAILERS HERE

RETAIL BUTCHER. (Pages 7 & 8.)

COOKING FATS RETAILER. (Pages 9 & 10.)

BUTTER & MARGARINE RETAILER. (Pages 11 & 12.)

BACON RETAILER. (Pages 13 & 14.)

SUGAR RETAILER. (Pages 15 & 16.)

How to use this Ration Book

INSTRUCTIONS TO PARENT OR GUARDIAN

Food was rationed during World War II and for several years afterwards.

The rationing during World War II was on a weight basis for the staple foods and on a points system for others. Points were transferable between some foods, which were therefore priced in money and in points. The amounts of the staples were so low that today's consumers would probably be reluctant to believe them. The cheese ration per adult per week for example was a mere

2 ounces (abbreviated oz.) (*i.e.* 57 g) and butter and margarine 6 oz. (170 g). Meat was on a price basis but adults were effectively allowed about 1 lb. (454 g) per week. The bacon and ham ration was 4 oz. (113 g).[13]

There was a determination that food supplies should be more secure in the post war period. The result was the 1947 Agriculture Act, modified in 1957, which enabled the government to become involved in pricing so that farmers could plan and invest with more confidence. The annual February price review between the Ministry of Agriculture and the National Farmers' Union (NFU) became a significant feature of the farming year. At the same time state funded agricultural research and advisory services were expanded and strengthened. The marketing and price support arrangements lasted until the 1970s, which was a time of increasing influence of both the European Community (EC) and of philosophies of free market economics. By contrast with the traditional agricultural products, some of the newer products like broiler chickens and many horticultural items have never been sold in anything but a free market.

The debate about the proper role of the state in agriculture still goes on, but now it is Europe wide. Under EC Agenda 2000 agricultural support is shifting from encouraging production to environmental support. In 1999 the UK Agriculture Minister, Nick Brown, announced the application of such a policy for the UK.

THE SCIENTIFIC ERA

Up to the early 18th century technical progress in England had been mainly farmer led, but in the 18th century agricultural science began to make its contribution. Although Arthur Young is generally considered to have been an agricultural journalist, his *Annals of Agriculture*, first published in 1784, was really a journal for the exchange of technical and scientific ideas, and it included contributions from many authors.[15] Some of the experiments reported seem quite modern, apart from the lack of statistical replication. For example John Symonds, Professor of Modern History at Cambridge, contributed an article on *A comparative experiment upon dung, lime, and compost, made in 1782, upon the lawn at St Edmunds Hill, Suffolk.* Some of the contributors reported work which was impressively fundamental. Samuel More wrote of *Experiments to ascertain the force necessary to draw various ploughs.*

The Board of Agriculture was established in 1793 by Pitt and was noted for having staged Sir Humphry Davy's lectures in 1803 on *"The connection of chemistry with vegetable physiology"*. The Board was wound up in 1822 but by 1838 the Agricultural Society of England was founded, to become the Royal Agricultural Society of England two years later.[26]

One of the best known early contributions of science to agriculture and food production was the introduction of superphosphate fertiliser by Lawes in 1842, but the *Journal of the Royal Agricultural Society* for the next few years was full of other impressive examples.

Experiment field plots. Photo: From the collections of the Department of Manuscripts and Special Collections, University of Nottingham.

The agricultural sciences both borrowed from the main stream sciences and made some significant contributions to them. In my view probably their most notable contribution to other disciplines was the development of the branch of mathematics now called statistics. This was developed at Rothamstead by people like Sir Ronald Fisher and Frank Yates for the analysis of field trials. It was necessary to separate the effects of the treatments under test in field plots from the effects of chance. To this day sampling techniques, analytical techniques, randomisation and replication form the basis of a good deal of research practice not only in food and agriculture but also in medicine, veterinary science, and industrial quality control world wide. Tables for statistical analysis were first published by Fisher and Yates in 1938.[60]

But statistics is not the only science to have felt the impact of the agricultural sciences. Many key findings in nutrition and biochemistry, now important in human nutrition, owe their origins to work on farm animals, and more recently the environmental sciences have benefited from the agricultural sciences.

It is probably not entirely coincidence that one of the greatest names in human nutrition, R.A. McCance, who, with E.M. Widdowson, did so much to place the feeding of the UK population onto a scientific footing during World War II rationing, originally intended to take an agricultural diploma. He switched to natural sciences, but worked for a while at the County Farm, Antrim in 1919, and described it as "valuable experience".[61] McCance and Widdowson's *The Composition of Foods,* first published in 1940, was published in its fifth edition in 1991 by a large team led by B. Holland and is still the standard reference text on the analysis of foods.[62]

AN EXAMPLE HISTORY OF A SCIENTIFIC DEVELOPMENT

An interesting example of a scientific topic relevant to both agriculture and to human nutrition, is the study of food energy. A brief digression on the history of the subject may be appropriate at this point. In these days of concerns about slimming and calories, and their interaction with exercise, such an example may be of interest.

In Derby Museum there is a collection of pictures by Joseph Wright (1734-1779). Having remained in Derby, rather than decamping to London as more ambitious artists of the time did, he was regarded as provincial and became known as Wright of Derby. Provincial or not his realistic style and his special use of light have given us some striking and memorable pictures. One of his specialities was depicting the excitement caused by the emerging sciences of the day, and from that genre a good example is not at Derby but in the National Gallery in London. *An Experiment on a Bird in the Air Pump,* painted in 1768, shows a group of onlookers marvelling at a demonstration of the 17th century discovery that animal life is dependent on air.

My own interest in quantitative food energy balance was presaged by a thought which crossed my mind during school dinners one day in about 1952. At a time of post-war austerity school dinners were seldom appetising, but on that day the offering must have been particularly unpleasant. I remember wondering, while struggling with a sample of limp, wet, boiled cabbage, of a very uncabbage-like colour, whether the energy required to lift it to my mouth would be exceeded by the energy it provided.

Long before my time the German chemist Liebig (1803-1873) had realised the connection between food and energy. This led the early English agricultural scientists to note the practical farming importance of warmth as equivalent to animal feed; for example in a communication by Playfair to the Royal Agricultural Society of England (RASE) in 1844.[63] English 19th century writers were, however, tempted to be a little condescending about continental contributions. Hyett wrote to the RASE in 1841 that "The useful work on organic chemistry by Professor Liebig, *though a foreigner* (my italics), is owing to the British Association for the Advancement of Science; but the Royal Agricultural Society opens much brighter prospects".[64]

From the realisation of the important association between food and energy has sprung much of our knowledge of how to feed ourselves and our farm animals. Until the connection between life and chemical heat generating processes was understood the subject of nutrition was bogged down in ideas which went back to the Greeks.[5,16] From the 5th century BC until the 17th century AD the theory of humours was used to explain life, and hence to make all sorts of strange and unfounded medical and nutritional recommendations. Life was supposed to be composed of blood, phlegm and two sorts of bile, corresponding to Aristotle's air, water, fire and earth, and diets were supposed to be adjusted to balance these according to the needs of the individual.

Sadly, despite our rich heritage of scientific nutrition it appears to be a feature of modern food obsessions that numerous current diets and fads still owe more to mystique than to objective biochemistry and physiology. After more than three centuries of inductive science we are still witnessing the deductive development of diets, owing more to the thoughts of their inventors than to the observation of facts and to accepted scientific principles. It seems that such intellectual back sliding is not new in food and agricultural matters. In 1841 Hyett complained that progress was being slowed because "As an art, the oldest of all and the most universal of all, it is still at this day more the work of empiricism, and less that of induction, than any other."[64]

As in many fields of study no one individual invented the whole edifice of the energy balances and the feeding of animals, but rather it developed gradually as a result of the efforts of many. Excellent and comprehensive

---

Footnote: Names, terms and units of measurement

Some terms and units may be of interest in a discussion of this particular aspect of the history of food and agriculture. Most people are aware of the connection between food and energy nowadays, and many are also aware of the term "calorie" because of an interest in diet and slimming. In international scientific work the calorie has been replaced by the joule (written J), which is a fundamental unit of energy named after James Joule (1818-1889) who realised that heat is a form of energy. Since 1 J per second is 1 Watt (W) the arithmetic is simple when making calculations ranging across biology and engineering. Because of the homely familiarity of the 100 W light bulb or the 1000 W (i.e. 1 kW) electric fire, it is easy to visualise the similarities as heat engines between, say, the 100 W lamp, the 150 W man or the 10 W chicken.

This particular field of study lacks a simple and elegant name, but if anything it is referred to as bioenergetics. The term appears to have been first used by Samuel Brody in his monumental work in 1945.[66] The term is precisely descriptive, meaning the energy of life, but perhaps not neat and elegant. Many of the measurements used in this subject are called calorimetry, which provides us with a memento of the calorie.

summaries of the early history have been provided in three landmark books on the subject, by Brody, Kleiber and Blaxter.[66,67,68,69] All three of these were themselves towering figures in this field and leaders of large teams of scientists. Some of the next few lines are based on such sources.

It had been realised by the 18th and early 19th centuries that animals burning oxygen gave out heat. In the mid 20th century numerous calorimetric measurements were made on farm animals, because a knowledge of their energy balance was of strategic value in feeding them, particularly during war time food shortages.[66,67,68,70,71,72]

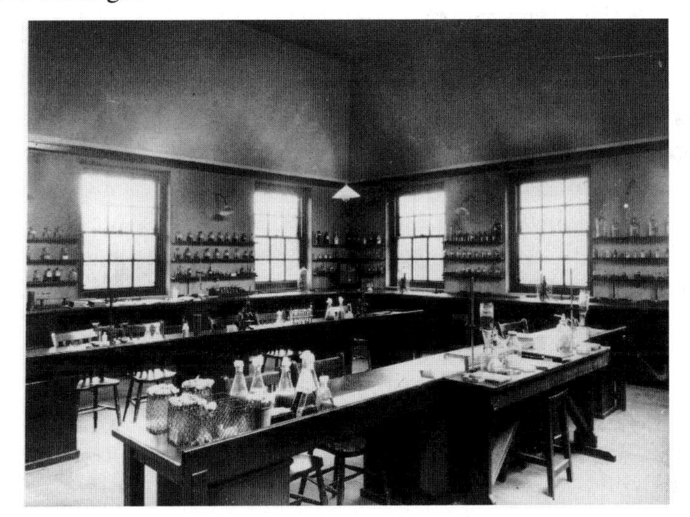

1924

1950

Agricultural laboratory work, 1924 and 1950. Photos: From the collections of the Department of Manuscripts and Special Collections, University of Nottingham.

Liebig's seminal realisation of the connection between food and heat, and the agricultural uptake of the idea by people like Playfair, provided the foundations for much of the later work on the scientific feeding of farm animals. It paved the way for the work of human nutrition pioneers like McCance and Widdowson.[62,63]

Early feeding systems for farm animals

Until the application of scientific methods the feeding of animals on the farm was a hit and miss business, based on the experience of the farmer and the stockman, and probably also based on making the best of what was available. Therefore much of the progress in animal husbandry was made by the farmers and breeders such as Robert Bakewell, and not by scientists and nutritionists. The great breeders knew a thing or two about feeding their animals, however, in order to exhibit the abilities of their breeds. Bakewell even ran his own feeding trials on the weights of turnips consumed by his sheep.[39]

Modern ideas on attempting to take chance into account in feeding trials were presaged by the Earl of Spencer in 1841 when he compared swedes and mangold-wurzels using two animals, but swapping the treatments from one animal to the other during the course of the experiment.[73]

Footnote: Units of energy in animal feeds

The Rev. W. Rham, in 1842, produced tables of equivalents of feedstuffs, by translating the work of a Monsieur Antoine of Nancy.[74] He took into account the important concept of a maintenance requirement, being that amount of food necessary to sustain the life of an animal or man without any gain or loss in weight and without producing anything. By 1886 Long was able to publish a table of feeding requirements for the growth of pigs (quoted by Wiseman).[75]

These important ideas became incorporated into later better known feeding systems, such as Oscar Kellner's starch equivalent system (1908), in which feeding stuffs were compared on the basis of their ability, relative to that of starch, to fatten cattle.[76] By the first two decades of the 20th century many American agricultural experimental stations were publishing recommended feeding regimes and maintenance allowances for dairy cows (quoted in Brody's book).[66]

Armsby's metabolisable energy and net energy system (published in 1922 but based on work he had done in 1903) gained wide acceptance, but in Europe the starch equivalent system remained in use until relatively recently, and was still the standard for ruminant feeding up to about 1960.[77] In Britain the metabolisable energy system was introduced into the Agricultural Research Council's *Nutrient Requirements of Farm Livestock* in 1965.[78]

Metabolisable energy, abbreviated ME, is that food energy which is left for use by the animal after faecal and urinary energy have been lost. Net energy (NE) allows for the energy used in digestion and absorption.

It is salutary to note that these relatively modern systems of feeding both farm animals and people were not the first attempts. The ancient Egyptians had the beginnings of such a technology thousands of years ahead of us. Their Pefsu value, sometimes translated as baking value, was the number of loaves or jugs of beer obtainable from 1 hekat (about 4.78 litres) of grain. Thus the Pefsu system permitted calculation of the equivalence of bread and beer. Rations for fattening geese were calculated in hekats of meal.[27]

Human nutrition and its agricultural heritage

McCance and Widdowson's *The Composition of Foods* (see above) is not the only aspect of human nutrition owing something to an agricultural heritage.[62,79] The Rowett Research Institute at Aberdeen, headed for many years by K.L.Blaxter as an agricultural centre, is now engaged in human nutrition under the directorship of W.P.T. James, co-editor of a series of textbooks on human nutrition.[80,81]

THE ERA OF CONSUMER SOVEREIGNTY

During the period from the outbreak of World War II until about the 1960s the progress and commercial development of food supply tended to be production led. But in recent years an abundance of supply and the declining importance of food in total consumer spending, combined with the advent of a globally competitive food market supported by air freight, has changed all that. Dr. John Beaumont, Chief Executive of the Institute of Grocery Distribution (IGD), in a paper to a BOCM PAULS poultry industry event in 1997, wrote of a change from "producer push to consumer pull".[82] The major retailers and the burgeoning food service industry are now the driving forces. An account of the history of agriculture from now onwards would not only be incomplete without reference to these influences, it would also be inappropriate and misleading.

Agriculture is part of the food industry. It always was, but it is now more directly so. This is surely a matter for pride rather than regret. The food industry, including food shopping and eating out, have become as much a part of British everyday life as driving on the left or the chimes of Big Ben. Food retailing must be one of the most finely tuned aspects of democracy ever devised. The retailers know very quickly what people really want, whatever they say they want, and are able to react and to offer it almost immediately.

All this emphasis on things which happen to food after it leaves the farm gate has meant that less and less of the money spent on food by the public finds its way into agriculture. It can be estimated from figures published by MAFF that in 1998 UK consumers spent about £71.5 billion on food and drink, though some estimates put the figure much higher.[11,83] Central Statistical Office and Euro PA figures for 1993 suggest £105.4 billion.[84] The gross output of British farms in 1999 was about £13.7 billion.[85]

It has not escaped the notice of agriculturalists that they might need to obtain a share in this added value. For the last thirty years or so it has been increasingly common for broiler chicken and turkey production companies, for example, to move into food further processing and food preparation.

Eating in the home is still important, but eating out accounts for more and more of the total food spend in Britain today.

This is nothing new. Probably one of the earliest commercial examples of this kind of enterprise was farmhouse cheese and butter making. Many dairy farms and estates of the 19th century and before had dairies. At first the processes used were varied, and therefore so were the products, but later standards were developed for named cheeses. A standard for farmhouse Cheddar was established in 1856 by Joseph Harding.[86]

An early added value enterprise: butter making in the farm dairy. This picture was taken at Sacrewell Farm and Country Centre, Peterborough.

Food production once concerned almost every rural householder. This domestic cheese press was in use until the early 20th century.

By the 1860s, following an outbreak of cattle plague and a consequent shortage of milk, it began to be difficult for individual farms to compete with imported cheese. Co-operative ventures sprang up to run cheese factories taking milk from a number of farms. Some of them are still in operation today producing their own speciality cheeses. In 1870 a Mr. Gilbert Murray displayed a model of the Derby factory of a co-operative at the Royal Show at Oxford.[87] Later the Milk Marketing Board (established in 1933) also manufactured milk products on behalf of producers.

More recently we have seen the emergence of farm shops and farm catering adding value to farm produce before it leaves the farm.

The food manufacturing industry is very old. The first manufactured foods were probably bread, oil and wine.[14] The fermentation by yeasts, which made bread and wine possible, was almost certainly an accidental discovery, perhaps about 4,500 BC. The basics of cheese making were also probably discovered unintentionally when milk was left standing too long. However these discoveries were the beginnings of the very influential business of adding value to food, preserving it, making it more portable, more tasty and more interesting. Food preservation was so important that the salt trade was international big business in classical times. It led to Roman interest in the Dead Sea and probably therefore to the conquest of Palestine, and thence to world Christianity through the Romans.[14] A Roman soldier's pay was called a salary, from the Latin word *sal* meaning salt, and salt is at the root of the word soldier.

The Arabs built the first sugar refinery at Candia in 1000 AD. Without preservation many foods quickly deteriorate and become useless, or even dangerous. Thus the invention of preservation by drying, salting, pickling, curing, smoking, spicing, sugaring (originally with honey), fermenting, potting and later freezing, canning and bottling, was another key to the feasibility of human survival and development, including travel, exploration and military activities.[88] The great geographical variety of different methods of preservation developed by different peoples of the world accounts for the rich profusion of flavours and regional specialities which we now enjoy. Most of the characteristics we associate with regional foods are due to the preservation methods and recipes, even the subtle differences between, say, the many cheeses of Britain. Some differences have acquired cultural significance.

It was military requirements which led to modern progress in food preservation and the foundations of the modern food industry. Napoleon needed preserved foods for his army's long expeditions. At that time Appert developed a technique of bottling after boiling, and then came cans made of welded tinplate. Pasteur realised why heating was needed to kill off the spoilage and food poisoning microorganisms, hence the term pasteurisation.[14] Later came cans of mild steel developed in England by B. Donkin in 1812, originally for travellers and explorers.[16] Surprisingly the earliest attempt at refrigeration was as early as 1859, by James Harrison. It was the availability of ice on board trawlers which led to the introduction of fish and chips in the 1860s and 1870s, probably as a development from the hot pie shop.[13]

Many of the modern types of food products, owing more of their money value to their cooking and processing than to the agricultural raw materials, date from the late 19th century. Margarine was first developed in the 1870s. Factory baking, as opposed to bread making in the home, came to Britain in the 1880s, though the millers of ancient Rome were really the first mass producers of factory made food. Interestingly the current interest in home bread making machines is bringing this piece of food history full circle. Factory made biscuits, jam and chocolate all date from the 1880s.[13,16]

This was a period which also saw the rise of the multiple retail grocers, which were eventually to dominate and to drive the supply and the nature of household food so thoroughly by the 1990s. The Equitable Pioneers of Rochdale started co-operative retailing of food in 1844 for social rather than commercial reasons, but the venture was a huge commercial success. Then came examples like J. Sainsbury (1869), Lipton's (1871), and the Home and Colonial (1888).[13] Self service stores started in USA in the 1930s and reached Britain in the late 1940s, though they did not become widespread until rather later.[59]

Despite the suspicion of the medieval civic authorities of dealers in food, by the 19th century the idea of farmers dealing direct with consumers was becoming impossible to sustain, though some examples survived for a long time. Indeed the notion of direct contact between consumers and producers was so valued that it ensured the survival of some urban farming enterprises for a remarkably long time. In the 1870s pigs were still to be seen in some London boroughs, and on the streets of some provincial towns until the 1920s.[89] A dairying text book published as recently as 1952 described town dairies, in which resident cows were milked in the town, and from which consumers bought milk direct.[90] Before 1865 there had been 550 cows in the Edgware Road and Paddington district of London.[91]

Towards the end of the 19th century consumers were already expecting food supply to include food preparation. By 1914 there were 25,000 fish and chip shops in England. J. Lyons opened tea shops in 1884 and there were 250 by the 1920s.[13] Canning and packaging did much to accelerate the growth of the taste for foods processed beyond their basic raw materials. Breakfast cereals appeared in the 1890s.[13]

An early example of moving cooking skills from the home to the factory was Bird's custard in the 1870s. The first health food was probably Hovis, invented by the Staffordshire baker Richard Smith in about 1885.[13] An early branded ready prepared sauce was HP, first made by a Nottingham grocer called F.G. Garton and then seriously marketed from 1903 by the Midland Vinegar Company of Birmingham.[92]

Fast food and eating out may seem like modern phenomena, and indeed they are very important and influential sectors of the modern food industry, but they are not as new as they seem. Apparently Marco Polo observed fast food shops and restaurants in 13th century China.[16] In 12th century London there were public cookshops to which citizens could take their own ingredients, and this innovation goes back to the time of Nebuchadnezzar in the near East, from whence it may have reached England *via* Spain and the Crusades. In 1690 a French visitor commented on the numerous cookshops of London.[93]

An early purpose designed English convenience food was the sandwich, named after John Montague, 4th Earl of Sandwich (1718-1792), who wanted something sustaining to take to round the clock gaming tables.[94] But this was not the first portable convenience food. Pies and pasties were traditional foods developed as convenient parcels for carrying meat when travelling,

An early example of moving cookery skills from the home kitchen to the factory. Reproduced from *Home and Country* with permission of the National Federation of Women's Institutes, WI Enterprises Ltd.[108]

mining or working in the fields. Later the hunting fraternity popularised the Melton Mowbray pork pie, which was found to be a handy snack to fit into a tunic pocket.[95]

For such reasons history has given Britain a rich heritage of traditional and regional foods. An inventory compiled in 1999 by Mason and Brown, as part of an EU project, listed 395 regional foods, including 36 cheeses. Yet they made the point that traditional regional foods are even more profuse in France, since Britain is more unified and some food traditions were lost during 13 years of rationing during and after World War II. Note that there are now far more regional and local speciality foods on the market than appear in the inventory classified as traditional. Mason and Brown defined traditional foods as having been made for three generations or longer.[96]

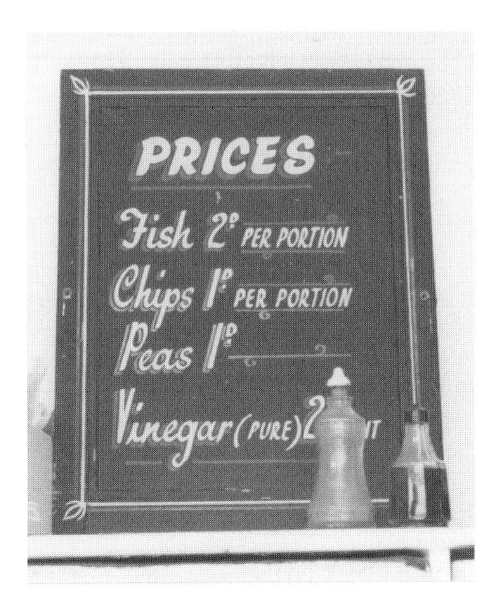

A mobile fish and chip shop of 1937. Photographed with permission of Leicester City Council, Leicester Museums.

Eating habits and food purchases have both led and been led by social changes. The growth of foreign holidays, and an immigrant population, gave rise to today's popularity of ethnic foods. These have done a great deal to advance the modern contribution of food to pleasures well beyond the mere satisfaction of hunger and the provision of nutrients. Italian restaurants came first, followed by Chinese and Indian.[59] A recent development in the enhancement of the pleasure of eating is the TV cook and celebrity chef, though the first TV cook, Marguerite Patten, was as early as the mid 1950s. Now there are sometimes 20 or so cookery or food related programmes per week on the five aerial TV channels and the BBC runs an annual food show at the National Exhibition Centre in Birmingham.

It is abundantly clear that the histories of food and of agriculture cannot really be treated as separate topics, nor could agricultural businesses prosper without realising that they are part of the food industry.

AGRICULTURAL EDUCATION, TRAINING AND RESEARCH

As we have seen the early innovators were commercial farmers. But they communicated with each other and an infrastructure was soon needed for the exchange of ideas. Thomas Coke of Holkham started the influential Holkham sheep shearings as farmers' discussion groups, as reported by Earl Spencer in 1842, thereby inventing a communications technique widely used much later by the National Agricultural Advisory Service (NAAS) and then by its successor, the Agricultural Development and Advisory Service (ADAS).[34]

Holkham Hall, Norfolk, the home of Thomas Coke (1754-1842). By permission of Holkham Estate.

Local agricultural shows became forums for ideas as well as for the exhibition of livestock improvements. As a national event the Royal Show, run by the Royal Agricultural Society of England (RASE), began in Oxford in 1839 and later settled on its permanent site at Stoneleigh, Warwickshire. The RASE assumed responsibility for the encouragement and dissemination of agricultural science and practical progress, adopting the motto "Practice with science".

A county instructor in dairying; an example from Staffordshire in the 1920s. The mobile exhibit was probably photographed during a visit to Uttoxeter.

By the mid 19th century agricultural education was under discussion. Some writers were advocating that village schools should teach gardening and agriculture.[53] As we shall see below, that was to happen in some areas, but in a slightly different way, as an extra-curricular function (e.g. Nottinghamshire District Agricultural Schools). But some were arguing for more than that. Charles Daubeny, Professor of Rural Economy at Oxford at the time, had studied agricultural education in France, Germany, Italy and Ireland, and offered his opinion that a check to the advancement of agriculture was "the want of a special education for the agricultural classes of all ranks and conditions".[97]

As a result of the belief held in the 1880s that education might be a palliative for the agricultural depression, an Act of 1890 gave county councils the powers to provide technical agricultural education. Originally the money came from the taxation of alcohol and became known as the whisky money. The responsibilities of the Board of Agriculture, set up in 1889 and later becoming MAFF, included determining the best means of developing educational facilities.

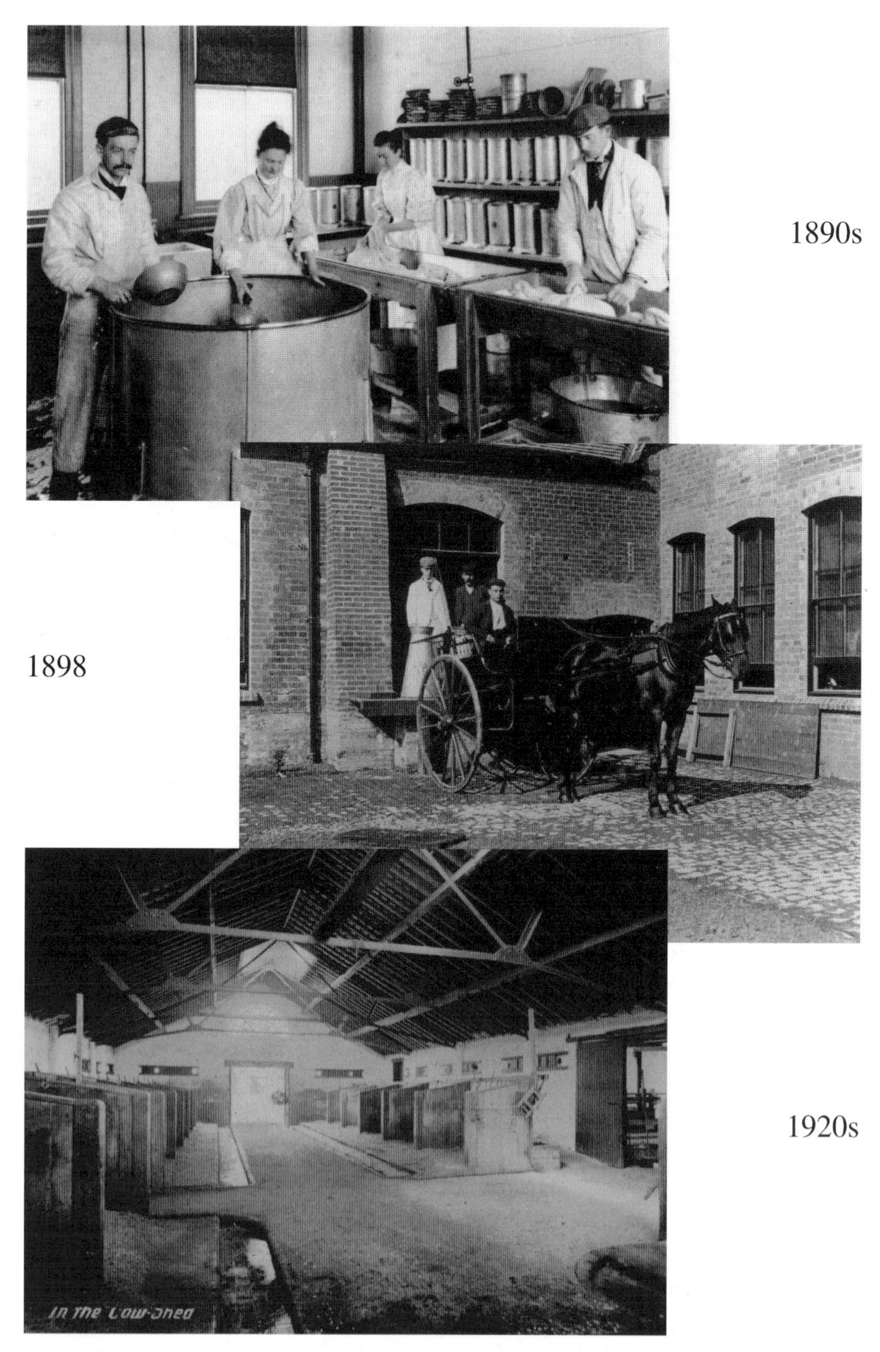

At the Midland Dairy Institute, later called the Midland Agricultural and Dairy College and then the Midland Agricultural College. Photos: From the collections of the Department of Manuscripts and Special Collections, University of Nottingham.

Eventually county councils established farm colleges, aimed mainly at the training of local people. Some agriculturally important counties such as Lincolnshire founded more than one college. Most county councils also had travelling instructors in subjects like dairying and poultry, based on the colleges, and the photo on page 41 shows an example from Staffordshire. Many counties even ran facilities like egg laying trials for the comparison of poultry breeds. Interestingly very few *countries* undertake this function any more, though there were national egg laying trials at Godalming, Surrey, until the early 1970s.

By combinations of the efforts of official bodies, local authorities and private benefactors, national colleges such as Harper Adams Agricultural College in Shropshire, Seale-Hayne in Devon and the Royal Agricultural College in Gloucestershire, were also developed to provide diploma level courses. Several universities offered degree and postgraduate courses in the agricultural sciences. Most of the agricultural colleges are now affiliated to universities.

Training in craft skills continued to be offered by the colleges and by specialist organisations like the Agricultural Training Board (ATB) until the present time, though with far less state funding in recent years, and the ATB was wound up in 1989. Nowadays there are numerous organisations offering training in craft and industrial skills, but they are very far from exclusively agricultural. For example first aid is taught by the Red Cross, the St.John's Ambulance and many others. Food hygiene and food handling courses are offered by many organisations and colleges.

Example college histories

The University of Nottingham, Sutton Bonington Campus, offers a fascinating example of the evolution of an institution originally benefiting from the pioneering educational initiatives of the 1890s. In 1955 Robinson recorded some of the history and in 1983 Tolley produced a more detailed paper on the subject, and it is mainly from these sources, as well as college annual reports and prospectuses, that the remarks below are taken.[98,99]

The Midland Dairy Institute grew out of a series of lectures organised by M.J.R. Dunstan for the University College Nottingham, whose Department of Agriculture had been established in 1892 under the Technical Instruction Grant of that year. Dunstan organised lectures and classes of instruction around the villages of the area, and local schools were encouraged to participate, some becoming Nottinghamshire District Agricultural Schools. Dairying was an important branch of farming in the area and a travelling dairy school became a key part of Dunstan's operations.

During 1893 discussions began between the University College Nottingham and the counties of Derbyshire, Leicestershire, Nottinghamshire and the Lindsey division of Lincolnshire. The report and prospectus of the Midland Dairy Institute and Agricultural Department for the year ended 31 March 1896 described the reasons for the choice of site acquired at Kingston, on Lord Belper's estate, in 1895. The location was considered both agriculturally suitable and central to the counties to be served.[100]

The University of Nottingham, Sutton Bonington Campus, in 1995.

By 1900 the Midland Agricultural and Dairy College had been built at Kingston. It taught milk production, butter making and marketing, and the making of at least seven English cheeses and seven continental cheeses. Among the important regional products whose technology and manufacture were taught was Stilton cheese, which had been made in the villages around Melton Mowbray long before it acquired its famous name in the 1740s. Its story has been told by Hickman.[101]

The construction of the buildings on the present site at Sutton Bonington began in 1910 but was interrupted by World War I, and the buildings were finally opened in 1928, having spent some years as a prisoner of war camp. The college later became the Midland Agricultural College and then in 1947 the Nottingham University School of Agriculture. It next became the Faculty of Agricultural and Food Sciences, having rightly greatly strengthened the emphasis on food to reflect the kind of trends in the food industry described above. A new food sciences building was opened by the Minister of Agriculture in 1997. In 1997 the Faculty merged with departments at the main Nottingham campus to become part of the Faculty of Biological Sciences. It is now the University of Nottingham School of Biosciences, Sutton Bonington Campus.

1940

Milking in 1940 at Sutton Bonington. This was typical of the hand milking of the time. Machine milking is now universal on commercial dairy farms.

2000

The University of Nottingham, Sutton Bonington Campus in 2000. In the background is a food sciences building reflecting the importance of the subject by that time.

Its library has made some of the research for this book possible. Its old students are still called Old Kingstonians after the original site of the College.

The educational emphasis at Sutton Bonington started from a tradition of practical agricultural subjects, gradually becoming more academic, and by the end of the 20th century preparing students for careers elsewhere in the food chain. This evolution reflected the changes reported earlier in this chapter.

As the bulk of the final value of food products has shifted from raw agricultural materials to today's prepared and added value products; as public concerns have changed; and as farm holding numbers have declined; so has the role of education changed. In their earlier days most of the colleges aimed to train farmers' sons and others as hands-on practitioners. By the mid 20th century at Sutton Bonington came diplomas and degrees for professionals and university status for the college. By the end of the 20th century came new disciplines.

Another example college, exemplifying these trends, and whose story was told by Heather Williams, was Harper Adams. Founded in 1901 for the practical instruction of Shropshire farmers' sons, by 2000 came university status and courses including food and consumer studies, business management, marketing, environmental issues and engineering. A Countryside Development Unit was introduced in 1999. Like Sutton Bonington, by the 20th century's end the college had become part of Britain's exporting sector by training overseas students and postgraduates.[102]

Harper Adams University College, now providing courses in rural, food and environmental subjects, was originally Harper Adams Agricultural College when founded in 1901 to provide training for farmers' sons.

Prof. Peter Wilson has documented the histories of several colleges which had received funds from the Frank Parkinson Agricultural Trust (established in 1943), of which he was Chairman. The colleges described included the Royal Agricultural College at Cirencester, which admitted its first students in 1845. This partly followed the advocacy of agricultural education by Prof. Daubeny. Wye College, in Kent, took students from 1894, but the institution was based on a much older non-agricultural college in buildings going back to medieval times. Wye is now part of the University of London. Seal-Hayne College, in Devon, started admitting students in 1920, though a farm had been purchased in 1907. The college is now attached to the University of Plymouth.[97,103]

Other colleges described by Wilson included Sparsholt College in Hampshire, originally dating from 1912. Hadlow College of Agriculture and Horticulture in Kent was formed by a combination of two earlier colleges: Sittingborne established in 1919 and the curiously named Horticultural College and Produce Company at Swanley, opened in 1889. Otley College of Agriculture and Horticulture in Suffolk started as such in 1970, but can trace its origins back to the Saxmundham Agricultural Trials of 1899. More recent English colleges discussed by Prof. Wilson were Hartpury College, Gloucestershire (1947), Berkshire Agricultural College (1949), Pershore College of Horticulture, Worcestershire (1954), Capel Manor College, Hertfordshire, (1968, but based on a medieval manor). Easton College, Norfolk, was established in 1974, but was based on an earlier farm institute.

In Scotland is the oldest agricultural college of them all, namely the Edinburgh School of Agriculture, which founded its first professorial chair of agriculture in 1790.[103]

Two recurring themes in agricultural education were firstly the surge of activity by founding fathers in the 1880s and 1890s, and secondly trends from practical agricultural studies towards more diverse and more academic approaches, in several cases through recent affiliation with universities. Having adapted to the changing needs of the times colleges of all the above categories now offer courses in a wide range of land based topics, not just agricultural subjects.

RESEARCH AND ADVISORY SERVICES

An example of the history of a research and advisory service is provided by the evolution of the National Agricultural Advisory Service (NAAS), which became the Agricultural Development and Advisory Service (ADAS) in 1971, and by the time of writing was ADAS Consulting Ltd.

The county councils had been providing instructors in subjects like poultry and dairying since the end of the first World War. In 1943 the Luxmore Committee reviewed methods of providing advice to farmers and recommended a national service under the Ministry of Agriculture. As a result NAAS was set up in 1946, having been briefly predated by the National Poultry Advisory Service which was established in 1945 and used as a pilot exercise.[104] By the 1970s the scale of the service was being reduced and in 1986 charges for its work were introduced. In 1992 ADAS became an agency owned by MAFF but not operating as a fully funded branch of it. In the spring of 1997 the organisation was offered for sale and was acquired by a management buy out, becoming ADAS Holdings Ltd.

Fundamental agricultural research was the responsibility of the Agricultural Research Council (ARC), later the Agriculture and Food Research Council (AFRC), and later still (after 1993) the Biotechnology and Biological Sciences Research Council (BBSRC). The Haldane Committee, set up by Lloyd George in 1917, began creating research councils and the ARC dates from 1931. It established 32 centres for the study of both problem based topics (e.g. grassland) and discipline based topics (e.g. animal physiology). Some of the centres were closed or amalgamated in the 1980s and 1990s.

NAAS and then ADAS also ran experimental farms, originally called Experimental Husbandry Farms and Experimental Horticultural Stations, and many of them are still operating in the new private company.

REFERENCES

1 Dossie, R. (1768) *Memoirs of Agriculture*. Nourse, London
2 Wiseman, J. (1986) *A History of the British Pig*. Duckworth, London
3 Davis, D. (1966) *A History of Shopping*. Routledge and Kegan Paul, London
4 Lane, M. (1995) *Jane Austen and Food*. Hambledon Press, London
5 Lauden, R. (2000) Birth of the modern diet. *Scientific American* **283:** 62-67
6 Ministry of Agriculture, Fisheries and Food (1995) *National Food Survey 1994*, HMSO, London
7 Ministry of Agriculture, Fisheries and Food (2000) *National Food Survey 1999*, The Stationery Office, London
8 Office for National Statistics (2000) *Family Spending*. The Stationery Office, London
9 Harvey, J. (1993) *Modern Economics*. Macmillan Press, Basingstoke
10 Blaxter, K.L. and Robertson, N. (1995) *From Dearth to Plenty. The Modern Revolution in Food Production*. Cambridge University Press

11 Ministry of Agriculture, Fisheries and Food (1999) *Agriculture in the United Kingdom 1998*, The Stationery Office, London
12 Spedding, C.R.W. (1996) *Agriculture and the Citizen*. Chapman and Hall, London
13 Burnett, J. (1989) *Plenty and Want. A Social History of Diet in England From 1815 to the Present Day*. Routledge, London
14 Toussaint-Samat, M. (1994) *History of Food*. Translated Bell, A., Blackwell, Oxford
15 Young, A. (1784) *Annals of Agriculture* **1** Goldney, London
16 Tannahill, R. (1988) *Food in History*. Penguin, London
17 Lattimore, R. (1959) *Hesiod*. University of Michigan Press
18 Fisher, N. (1998) Work and leisure. In: *The Cambridge Illustrated History of Ancient Greece*. Edit. Cartledge, P., Cambridge University Press, 193-218
19 Columella, L.J.M. translated into English 1745. *Of Husbandry in Twelve Books and His Book Concerning Trees*. Millar, London
20 Lodge, B. (1873) Translation of *Palladius on Husbondrie*. Early English Text Society, London
21 Peters, M. (1771) *The Rational Farmer: Or a Treatise on Agriculture and Tillage*. 2nd edition, Flexney, London
22 Mills, J. (1762) *A new and complete system of practical husbandry*. Baldwin, London
23 Soffe, R.J. (1995) *The Agricultural Notebook*. Blackwell Science Ltd., Oxford
24 Spedding, C.R.W. (1992) *Fream's Principles of Food and Agriculture*. 17th edition of *Fream's Elements of Agriculture*. Blackwell Scientific Publications, Oxford
25 Watson, J.A.S. and West, W.J. (1956) *Agriculture. The Science and Practice of British Farming*. 10th edition. Oliver and Boyd, Edinburgh
26 Prothero, R.E. (Lord Ernle)(1936) *English Farming Past and Present*. 1961 edition, with introductions by Fussell, G.E. and McGregor, O.R., Heinemann and Frank Cass and Co., London
27 Kemp, B.J. (1989) *Ancient Egypt*. Routledge, London
28 Bell, H.I. (1948) *Egypt*. Clarendon press, Oxford
29 Rundle, J.R. (1955) Laxton today. *Agriculture* **62:** 170-172
30 Beckett, J. (1989) *Laxton - England's Last Open Field Village*. Trustees of the Laxton Visitor Centre
31 Humberside Archaeology Unit (1989) *The Isle of Axholme. Man and Landscape*. Humberside County Council
32 Atkins, J., Hammond, B. and Roper, P. (1999) *A Village Transformed. Keyworth 1750-1850*. Keyworth and District Local History Society

33 Wade Martins, S. (1990) *Turnip Townshend.* Poppyland Publishing, North Walsham

34 Spencer, the Earl (1842) On the improvements which have taken place in West Norfolk. *Journal of the Royal Agricultural Society of England* **3:** 1-9

35 Tull, J. (1762) *Horse-hoeing Husbandry.* Millar, London

36 Duhamel, M. (1762) *Duhamel's Husbandry. A Practical Treatise on Agriculture.* Hitch *et al.*, London

37 Pawson, H.C. (1957) *Robert Bakewell: Pioneer Livestock Breeder.* Crosby Lockwood, London

38 Fraser, A. (1959) *Beef Cattle Husbandry.* Crosby Lockwood, London

39 Stanley, P. (1995) *Robert Bakewell and the Longhorn Breed of Cattle.* Farming Press Books, Ipswich

40 Griffiths, E. (1998) Sir Henry Hobart: a new hero of Norfolk agriculture. *Agricultural History Review* **46:** 15-34

41 Scott Watson, J.A. and Hobbs, M. E. (1937) *Great Farmers.* Selwyn and Blount, London

42 Pusey, P. (1851) On Mr. M'Cormick's reaping machine. *Journal of the Royal Agricultural Society of England* **12:** 160

43 Gammon, R. and Fairhead, E. (1990) *One Man's Furrow.* Webb and Bower, Exeter

44 Culpin, C. (1992) *Farm Machinery.* Blackwell Scientific Publications, Oxford

45 Winter, M. (1995) *Harry Ferguson and I.* Winter

46 Barker, J. (2000) *Wordsworth, a Life.* Viking, Penguin, London

47 Marks, H.E. and Britton, D.K. (1989) *A Hundred Years of British Food and Farming.* Taylor and Francis, London

48 Ministry of Agriculture, Fisheries and Food (1989) *Agriculture in the United Kingdom 1988*, HMSO, London

49 Ministry of Agriculture, Fisheries and Food (1994) *Agriculture in the United Kingdom 1993*, HMSO, London

50 Tomalin, D. and Hanworth, R. (1998) *House For All Seasons. A Guide to the Roman Villa at Brading, Isle of Wight*, Oglander Roman Trust

51 Vaughan, A. (1997) *Railwaymen, Politics and Money.* Murray, London

52 Dexter, K. and Barber, D. (1961) *Farming for Profits.* Penguin, Harmondsworth

53 Baxter, J. (1846) *Baxter's Library of Agriculture.* First published 1830. Lewis, Sussex

54 Perren, R. (1995) *Agriculture in Depression, 1870-1940*, Cambridge University Press

55 Jackson, D. and Russell, O. (1987) *The Great Central in LNER days. 2.* Ian Allen, Shepperton

56 Taylor, A.J.P. (1965) *English History 1914-1945*. Oxford University Press
57 Zuckerman, L. (1998) *The Potato*. Macmillan, London
58 Goodenough, S. (1997) *Jam and Jerusalem*. Collins, London, and National Federation of Women's Institues
59 Hardyment, C. (1995) *Slice of Life. The British Way of Eating Since 1945*. BBC Books
60 Fisher, R.A. and Yates, F. (1963) *Statistical Tables*. 6th edition. Oliver and Boyd, Edinburgh
61 Ashwell, M. (1993) *McCance and Widdowson - A Scientific Partnership of 60 years 1933 to 1993*. British Nutrition Foundation, London
62 Holland, B., Welch, A.A., Unwin, I.D., Buss, D.H., Paul, A.A. and Southgate, D.A.T. (1991) *McCance and Widdowson's The Composition of Foods*. The Royal Society of Chemistry and Ministry of Agriculture, Fisheries and Food
63 Playfair, L. (1844) On the general principles of nutrition and on the food intake of cattle. *Journal of the Royal Agricultural Society of England* **4:** 215-237
64 Hyett, W.H. (1841) Experiment on the application of nitrate of soda as a manure for wheat. *Journal of the Royal Agricultural Society of England* **2:** 139-146
65 Smith, A. (1776) *The Wealth of Nations*
66 Brody, S. (1945) *Bioenergetics and Growth*. Reinhold, New York
67 Kleiber, M. (1961) *The Fire of Life*. Wiley, New York
68 Kleiber, M. (1975) *The Fire of Life*, 2nd edition, Krieger, New York
69 Blaxter, K.L. (1989) *Energy Metabolism in Animals and Man*. Cambridge University Press
70 Mitchell, H.H. and Kelley, M.A.R. (1933) Estimated data on the energy, gaseous and water metabolism of poultry for use in planning the ventilation of poultry houses. *Journal of Agricultural Research* **47:** 735-745
71 Findlay, J.D. (1950) The effects of temperature, humidity, air movement, and solar radiation on the behaviour and physiology of cattle and other farm animals. *Bulletin Hannah Dairy Research Institute* **9**
72 Lundy, H., MacLeod, M.G. and Jewitt, T.R. (1978) An automated multi-calorimeter system: preliminary experiments on laying hens. *British Poultry Science* **19:**173-186
73 Spencer, the Earl (1841) On the comparative feeding properties of mangold-wurzels and Swedish turnips. *Journal of the Royal Agricultural Society of England* **2:** 296-298
74 Rham, W.L. (1842) On the comparative value of different kinds of fodder. *Journal of the Royal Agricultural Society of England* **3:** 78-82

75 Wiseman, J. (2000) *The Pig. A British history*. Duckworth, London
76 Kellner, O. (1908) Translated by Goodwin, W. (1926) *The Scientific Feeding of Animals*. Duckworth, London
77 Armsby, H.P. (1922) *Nutrition of Farm Animals*. Macmillan, New York
78 McDonald, P., Edwards, R.A. and Greenhalgh, J.F.D, (1988) *Animal Nutrition*. 4th edition. Longman Scientific and Technical, Harlow
79 McCance, R.A. and Widdowson, E.M. (1940) *The Chemical Composition of Foods*. Medical Research Council Special Report No. 235
80 Smith, D. (1998) The Agricultural Research Association, the Development Fund, and the origins of the Rowett Rearch Institute. *Agricultural History Review* **46:** 47-63
81 Garrow, J.S., James, W.P.T. and Ralph, A. (2000) *Human Nutrition and Dietetics,* 10th edition, Churchill Livingstone
82 Beaumont, J. (1997) *The Consumer Has Spoken*. BOCM PAULS conference, Edit. Cessford, J., Wishaw
83 Ministry of Agriculture, Fisheries and Food (1999) *National Food Survey 1998*, The Stationery Office, London
84 Strak, J. and Morgan, W. (1995) *The UK Food and Drink Industry*. Euro PA and Associates, Cambridge
85 Ministry of Agriculture, Fisheries and Food (2000) *Agriculture in the United Kingdom 1999*, The Stationery Office, London
86 Ingram, A. (1991) *Dairying Bygones*. Shire Publications, Princes Risborough
87 Davis, J.G. (1965) *Cheese,* J. and A. Churchill Ltd., London
88 Shephard, S. (2000) *Pickled, Potted and Canned. The Story of Food Preserving*. Headland, London
89 Malcolmson, R. and Mastoris, S. (1998) *The English pig. A history*. Hambledon Press, London
90 Russell, K.N. (1952) *Fishwick's Dairy Farming*. Crosby Lockwood, London
91 Stapledon, R.G. (1944) *The Land Now and Tomorrow*. Faber and Faber, London
92 Landen, D. and Daniel, J. (1985) *The True Story of HP Sauce*. Methuen
93 Paston-Williams, S. (1993) *The Art of Dining*. National Trust Enterprises, London
94 Crystal, D. (1990) *The Cambridge Encyclopedia*. Cambridge University Press
95 Hickman, T. (1997) *History of the Melton Mowbray Pork Pie*. Alan Sutton Publishing, Stroud
96 Mason, L. and Brown, C. (1999) *Traditional Foods of Britain. An Inventory*. Prospect Books, Totnes
97 Daubeny, C. (1842) On the public institutions for the advancement of agricultural science which exist in other countries, and on the plans

which have been set on foot by individuals with a similar intent in our own. *Journal of the Royal Agricultural Society of England* **3:** 364-386

98 Robinson, H.G. (1955) Sutton Bonington. *Agriculture* **62:** 180-183

99 Tolley, B.H. (1983) M.J.R. Dunstan and the first Department of Agriculture at University College, Nottingham, 1890-1900, *Transactions of the Thoroton Society of Nottinghamshire*

100 Midland Dairy Institute and Agricultural Department reports and prospectuses for 1896-1901

101 Hickman, T. (1996) *The History of Stilton Cheese*. Alan Sutton Publishing, Stroud

102 Williams, H. (2001) *The Lure of the Land. A Century of Education at Harper Adams*. Harper Adams University College

103 Wilson, P. (2000) *Tale of Two Trusts*. The Memoir Club, Spennymoor

104 Hewson, P.F.S. (1986) Origin and development of the British poultry industry: the first hundred years. *British Poultry Science* **27:** 525-540

105 Arch, J. (1966) *The Autobiography of Joseph Arch*. Edit. O'Leary, J.G., MacGibbon and Kee, London, First published 1898 as *Joseph Arch: The Story of His Life*

106 Horn, P.L.R. (1971) *Joseph Arch*. Roundwood Press, Kineton, Warwicks.

107 Cobbett, W. (1830) *Rural Rides*. 1967 edition with introduction by Woodcock, G., Penguin, Harmondsworth

108 Kitchen, P. (1990) *For Home and Country*. Ebury Press, London

109 Andrews, G.H. (1853) *Modern Husbandry*. Nathaniel Cooke, London

110 Youatt, W. (1834) *Cattle; Their Breeds, Management and Diseases*. London

111 Lyth, P. (1989) *A History of Nottinghamshire Farming*. Cromwell Press, Newark

112 Dearlove, P. (2000) Fen Drayton, Cambridgeshire: An Estate of the Land Settlement Association. In: *Rural England*. Edit. Thirsk, J., Oxford University Press

113 Loyn, H.R. (1962) *Anglo-Saxon England and the Norman Conquest*. Longman, London

Stone age tools. Photographed with permission of the Farm and Country Centre, Sacrewell.

The modern countryside reflects its past.

A threshing scene re-enactment at an agricultural show.

Cattle at an agricultural show in the Midlands.

A Fordson tractor at work in 1939.

Dairy work, 1950. Photos reproduced with pemission of the Department of Manuscripts and Special Collections, University of Nottingham.

Vintage tractors at an enthusiasts' event.

In parts of Britain from the 1990s onwards the countryside is much more than a place for food production. This is Sawrey, the home of Beatrix Potter, in the Lake District. The beef cattle have their place in the food supply chain, as have the Herdwick sheep which she encouraged and bred, but visitors see them as part of the scenery.

Rapeseed crops are brightly coloured and conspicuous at certain times of the year.
Photo: DJA Cole

A wild flower farm. With permission of Naturescape, Langar, Nottinghamshire.

Peas: an example of a crop which contributes to crop rotations by fixing atmospheric nitrogen.[51]

A modern farm shop at Middle Farm, Firle, Sussex. Photo: John Pile
©2000 Middle Farm Limited.

A modern supermarket. Photo: DJA Cole.

A farmers' market. Photo: © Countryside Agency, Mark Osbourne.

Modern food retailing is often on a large scale. Photo: Tesco.

2

FROM HUNTER GATHERER TO FARMER

Ere virgin earth first feel th'invading share,
The genius of the place demands thy care,
The culture, clime, the winds and changeful skies,
And what each region bears, and what denies.

Virgil (70-19 BC), quoted by Baxter (1846)[1]

THE DEVELOPMENT OF ANIMAL HUSBANDRY: A SOCIAL REVOLUTION

The events described in Chapter 1 occurred during the period covered by written historical records, but two key revolutionary events took place much earlier. Without these two events history would not have happened, and that means all history, not just food and agricultural history. The first of these dramatic events was the domestication of farm animals and the second was the invention of cereal farming.

It may have been about 20 million years ago when our primate ancestors abandoned their vegetarian and fruit eating existence and replaced it with hunting and gathering.[2] This probably occurred when climate changes shrank the forests. According to some authorities from about 2 million years ago small game and fish may have been the main prey, followed by big game by about 500,000 years ago.[3]

It has been claimed that the food manufacturing industry dates from about 2 million years ago, when *Homo habilis* first use tools to prepare food and discovered that meat kept better in cool caves than it did in the open.[4]

About 800,000 years ago the use of fire must have greatly improved the palatability of such a diet, though the invention of an oven had to wait till about 30,000 years ago. In the Palaeolithic era cooking involved putting hot stones into skins containing the meat. The ground ovens of the Celts were also based on hot stones, though cauldrons seem to have been developed at about the same time.

Grains, seeds and fruits were gathered, probably by the women and children. These would have balanced the meat hunted by the bands of men, and provided a mix not too unlike modern meat and two vegetables menus. Had the diet not been suitable our ancestors would not have survived. Since meat diets are more concentrated in some nutrients than most plant materials more time was available, after exclusively herbivorous diets were abandoned, for other pursuits such as devising tools.[2] But none of this gathering and collecting counts as agriculture.

A stone age tool. Photographed at Sacrewell Farm and Country Centre, Peterborough, with permission.

The agricultural revolution, for that is what it was, may have come quickly when it arrived. It may have taken only a few centuries to make the switch from hunter gathering to farming.[5] However, until recent times domestication and the development of breeds were probably slow processes.[6] Originally there was no control over breeding, and in his book on animal breeding Hagedoorn pointed out in 1954 that there would have been a considerable amount of hybridisation between domestic and wild types.[7] Later the domesticated animals were separated from the wild types, and later still selective breeding for desirable traits occurred.

What motivated such a drastic change from hunting animals to farming them? There are many opinions, but a view recently put forward is that since meat did not keep very well in those pre-refrigerator times, the herding of animals may merely have been a method of preserving meat on the hoof.[8]

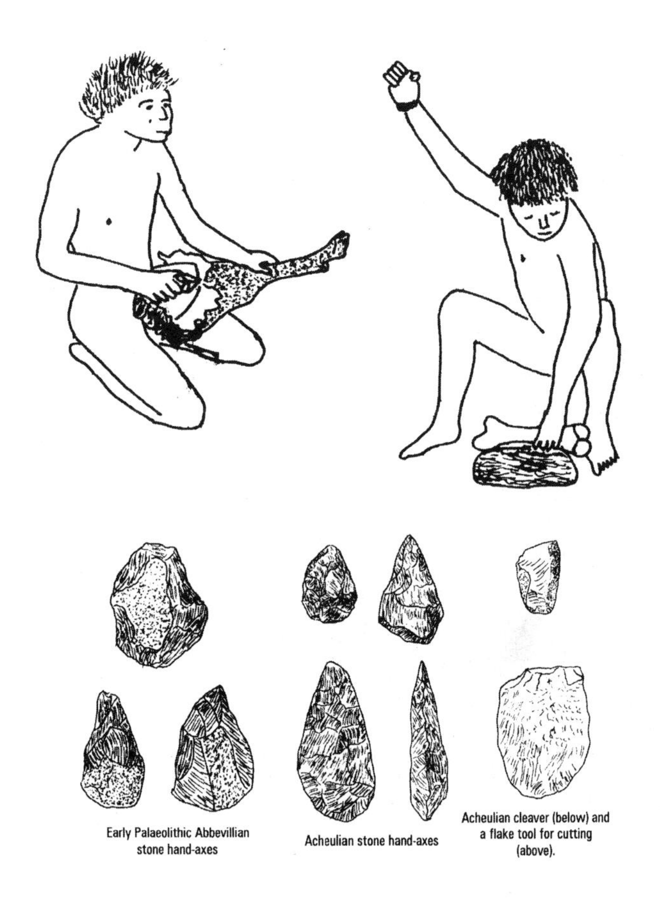

Stone age scenes and tools. Drawings from Rixson, reproduced with permission.[13]

The precise mechanism of the process of domestication of farm animals is not clear. One suggestion is that early man had driven game into funnel shaped traps and that it was a short step from this to closing the open end of the funnel to make a pen or a field. Another possibility is that hunters sometimes brought back to the camp the orphaned young of their prey as pets for the women and children or as hunting decoys.[9] Sheep may have been domesticated after the hunters noticed the ability of dogs to round them up, though there is some doubt if the wild progenitors were that easily rounded up.

The ancestors of our cattle and sheep may originally have been unwelcome visitors which found that growing crops provided easy pickings. Then it dawned on people that it was easier to catch them than to chase them away. Yet another possibility is that early man lured cattle with salt licks and then captured the young to tame them.[10,11,12] This would have been safer than capturing adult wild aurochs, the bulls of which were massive and armed with huge horns.[13] In the case of cattle there may even have been a fifth possible route to domestication. They may have been penned for religious sacrifice because their horns resembled a crescent moon. Penned animals may then have mated. Derrick Rixson made the point that early man probably scavenged carcasses before he could bring them down.[13]

Hunted animals. Drawings from Rixson.[13]

However domestication happened it was revolutionary because it led to a settled life with permanent homes and did away with the need to follow the game herds. Settlement paved the way for the next, and even more consequential revolution, namely crop production, which is the topic of the next section of this chapter. The animal husbandry revolution was cultural as well as economic, and this is another reason to suppose that it may have been slow. In hunter gathering cultures great hunters are the heroes. It must have taken some considerable adjustment to the male psychology to adapt to the different mentality of acquiring status by conserving the animals rather than by hunting them.[14] Thus a number of authorities considered that women played a significant part.[9]

The early domesticators could not have foreseen the final consequences of their endeavours, and therefore were operating without fully knowing what they were doing. They knew only of the immediate uses for the animals (meat, skins and bones for tools) which had already been developed from hunting. They could not have guessed at the secondary uses later put to farmed animals, such as milk, wool, motive power, war, sport, currency and prestige.[14] Meat was not the only contribution of butchery to early man's welfare. The non-food uses of parts of animal carcasses were very important to the developing cultures of the time. Thus Rixson tabulated an astonishing 38 crafts which had become dependent, or partly dependent, upon carcasses by medieval times.[13]

Early husbandry. From Rixson.[13]

The animals which became domesticated have certain characteristics in common. They are herding and flocking species, naturally given to sticking together in fairly close groups. This made it easier to trap and then to enclose them. Solitary creatures, or animals preferring widely spread groups, could not be kept in herds or flocks without serious operational problems. The domesticated species were relatively docile, and taming may have merged indistinguishably into domestication. Species readily imprinted on human substitute parents lent themselves to the process.[9,14]

The early domesticated species were the ruminants, able to utilise grass and rough herbage and thus greatly expanding the resources available for man's exploitation. The non-ruminant exception was the omnivorous pig, which may have been attracted to the camps in search of scraps and was probably a pest at first, just as escaped feral pigs have occasionally become pests in modern times. Species for domestication also had to be good mothers, able to reliably raise their young in their new circumstances. Among the smaller animals the chicken is discussed separately later, because it played no essential part in the development of settled agriculture.

Dosing a sheep, from Baxters' *Library of Agriculture*, 1846.[1]

The first domestication was probably that of the sheep (*Ovis aries*) about 11,000 years ago, somewhere in the middle east in the area which is now Iraq.[15] It may have reached Britain by about 4,000 BC. There is debate about whether the sheep as we know it is derived from one or more wild types, contenders being the Urial and the Mouflon. Some have suggested that wool types came from the Urial, through Soay type breeds, while hair sheep came from the Mouflon. Chromosome studies suggest that the Mouflon is the more likely ancestor, though modern sheep are so diverse that it is probably wrong to look for affinities between breed types and modern forms.[9] The Roman agricultural writer Palladius appears to have been unimpressed by the intelligence of sheep. His advice was that "Ewes should have large bones and long soft fleeces. Their pasture should be rich, and free from briars, which would unclothe the silly innocents, and also tear their skin", (from the translation by the Rev. Barton Lodge in1873).[16]

The goat (*Capra hircus*) came soon afterwards in the same region sometime before 7,000 BC.[17]

The pig (*Sus scrofa*) became domesticated by about 7,000 BC in Anatolia and at several other sites. Its ancestor, the wild boar, roamed widely in Europe.[18] The pig is an animal which seems to inspire fascination apart from its food value. It occurs in children's stories and nursery rhymes, and at least three histories of pigs in Britain have been popular in the last few years.[19,20,21] It was a stalwart component of English village economies for long periods of history, and of cottage economies until well into the 20th century. It features in modern agriculture in both the intensive production of commodity pork and bacon and in the new niche markets. One of today's most popular sandwiches is the BLT (bacon, lettuce and tomato).

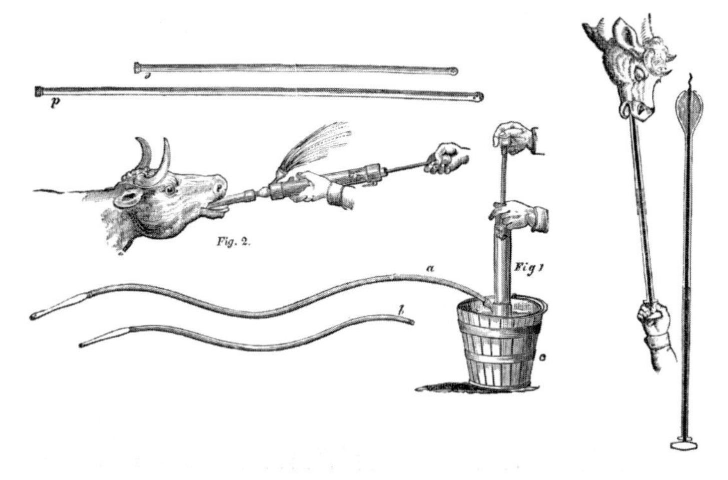

Relieving cattle choked with turnips. From Baxters' *Library of Agriculture.*[1]

Cattle (*Bos taurus*) were probably domesticated a little later, being larger and having rather dangerous males, by about 6,000 BC.[22] All modern European cattle are believed to be derived from the wild aurochs (*Bos primigenius*). Cattle probably reached Britain by the Bronze Age about 1,000 BC. The wild aurochs was wide ranging in Europe and Asia, but sadly is now extinct, the last cow having died in Poland in 1627. The Zebu, or Brahman, is *Bos indicus,* thought to originate from India, and some cross breeds used in hot parts of the world have *Bos indicus* contributing to their ancestry, conferring heat tolerance.[23]

However recent research based on mitochondrial DNA analysis, has raised new questions about the origins of modern cattle. Evidence has been found supporting the idea that domestication occurred more than once. Modern European cattle were probably not domesticated within Europe, despite that fact that the aurochs ranged widely, including in Britain, but are of near-Eastern origin. A separate African domestication probably also took place. DNA evidence suggests that *Bos taurus* and *Bos indicus* separated genetically some 100,000 years ago.[24]

The rabbit *(Oryctolagus cunuculus)* was probably domesticated about 3,000 years ago in Spain and perhaps earlier in Turkey. It reached Britain with the Normans, who established rabbit gardens under the care of warreners. Originally it was the young which were called rabbits while the older animals were coneys.[25]

A New Leicester bull.

A Shorthorn bull.

A Hereford cow.

Engravings from Youatt (1834)[46] collated by Rixson.[13]

Modern south Devon cattle.

Vertebrates are not the only animals which have been domesticated. There is, of course, an important insect, which provided a sweetener a long time before sugar was available. The honey of wild bees was collected in Neolithic times. By the 7th century BC the Egyptians were keeping bees (*Apis mellifera*) in hives.[3]

Not only did livestock husbandry make settled life possible but also its importance was reflected in its effects on cultural, religious and commercial developments. Cattle were wealth, both literally and symbolically. In some cultures they were used as currency. Cheeses made of sheep's and goat's milk were among the earliest preserved foods, with all the advantages and security that preservation brings. Developing languages had many words for the animals and for cheese. Animals had been in art since the caves of pre-historic man, and domesticated livestock occurred widely in the art of Egypt and of the classical Greek and Roman periods. They played significant parts in the religions and mythologies of both prehistoric and historic man.

Between the completion of domestication and the beginning of recorded history there are a few uncertainties. One revolves around the question of the survival through the northern European winter of the early ruminant farm animals. The traditional opinion was that in autumn all except important breeding stock were slaughtered, but there are some doubts about the arithmetic viability of that proposition.[9] There is not much doubt that many of the less fit animals would have died or had to be culled, but that may not have been deliberate. Hay making may have been copied to a limited extent from the natural habits of the hare. In Iron Age Britain it is possible that cattle and sheep were fed brushwood during the winter. The early breeds may have been better able to utilise such rough fodder than are the modern breeds. Satisfactory answers to the question of pre-historic winter feeding are not readily available.

*Messrs. Bond and Harwood's Linseed Crusher, Ipswich, Suffolk.**

A linseed cake crusher. From Baxters' *Library of Agriculture*, 1846.[1]

A more recent linseed cake crusher, circa 1940, photographed at Sacrewell Farm and Country Centre.

For several thousand years changes in the genetic constitution of the livestock species were slow. Useful traits were gradually enhanced, by deliberate or accidental selective breeding. Numerous local breeds evolved through geographical isolation, and this has given us a rich and varied heritage. However systematic breeding had to wait for the era of Robert Bakewell (1725-1795), the Colling brothers and Coates. Breeds acquired enthusiasts and breed societies, and in 1822 Coates compiled his famous Shorthorn herdbook. Agricultural shows became focal points for exhibiting stock.

Robert Bakewell of Dishley, Loughborough (1725-1795). Portrait reproduced by courtesy of the Royal Agricultural Society of England.

By the late 1950s there was some criticism of the notion of showing for breed type conformation rather than for objectively measuring performance, though by that time objective testing schemes were also in full swing, particularly for dairy cows. In the 1990s the need for niche markets has given a new role to some of the local breeds, and there are even butchers specialising in rare breed meats. The Rare Breeds Survival Trust encourages such activities. Breed societies for the rare breeds are active and some have documented the histories of their breeds.[26] Perhaps we have come full circle and the selection and showing for breed type is finally paying off. Ironically by correctly selecting for what consumers wanted at the time, the great improvers of the 18th and 19th centuries produced animals too fat for modern tastes. The Durham Ox, bred by Charles Colling (1750-1836), which toured on show around 1801, weighed in at 3,024 lbs (1,372 kg).[27]

In recent history farmers and animal breeders, supported by, for example, improvements in animal nutrition, have continued to improve yields. Table 2.1 gives some examples from published sources.

TABLE 2.1 CONTINUING PROGRESS IN THE PERFORMANCE OF FARM ANIMALS

	Years	*Approximate % increase per year*
Milk yield (litres per cow per year)	1989/91 to 1999	1.9
Piglets reared per sow per year	1990 to 1999	0.5
Eggs per hen per year	1988/89 to 1999	0.7

Sources: DEFRA (2001)[48]
Close and Cole (2000)[49]
Data from Table 3.4, Chapter 3

Interestingly there seems to be a fairly commonly held view, in these food conscious times, that our natural diet is vegetarian. If the events chronicled in this chapter are anything to go by, and if the early parts of the time scales cited above and in the chronology are anything like the truth, then it seems that for a long time we have been omnivors, not herbivors. This could make a difference to some of today's food and nutrition debates. In the 1860s anaemia was common amongst adolescent working class girls, and it sometimes caused a pallor so severe that it was called the green sickness. It may have been due to a lack of meat, coupled with lack of calcium.[2]

Inspecting a bull in 1940. Photo: From the collections of the Department of Manuscripts and Special Collections, University of Nottingham.

A modern dairy cow, bred for high yields and efficient production. Photo: A Lock.

GRASSES, CROPS AND CIVILISATION

The grasses and cereals (botanical family *Gramineae*) play a significant part in the appearance of what we have come to think of as typical British countryside. But their historical significance runs deeper than that. Western civilisation, and the development of life as we know it, depended upon their domestication and exploitation. The rest of history, the arts, science, technology and modern comforts followed. Because their domestication was such a consequential event many historians, archaeologists and botanists have called it the Neolithic revolution.[11] The notion of the historical importance of the domestication of the cereal crops is not new; it was included in agricultural botany courses of the 1950s and before.

Ryegrass for cutting.

Early man probably ate grass seeds for nearly a million years. Such carbohydrate rich seeds, gathered and brought back to the camp, would have been a useful supplement to the big and small game discussed above. Around human encampments there must have been many occasions when the ground became trodden and disturbed. The seeds of many of the grasses germinate well in disturbed soil, and it was probably an accidental discovery that spilled grains gave rise to crops of the progenitors of the cereals. About 9,000 BC this inspired some visionary Neolithic innovators to have the revolutionary idea that it might be easier to deliberately plant the grains rather than to go out searching for them. On the evidence of DNA finger printing, a group of lines of einkorn (*Triticum monococcum boeticum*) from southeast Turkey has been proposed as the progenitor of cultivated einkorn. This was one of eight founder crops launched in that area at about that time.[5,28,42]

Cereal harvesting in 1940.

A reaper and binder, 1940. Both pictures from the collections of the Department of Manuscripts and Special Collections, University of Nottingham.

Rice (*Oryza sativa*) was probably domesticated later, though the exact date is uncertain. It may have been about 3,000 BC in India; but perhaps in China somewhat earlier, by 5,000 BC.[3,29]

The cultivation of the cereals offered the right conditions for several dramatic and consequential changes. The availability of important new food commodities led to the end of the dependence on herding and shepherding on the move, and permitted the development of farming and permanent settlements as later centuries came to know them. Groups of people could settle and survive without the entire population being food providers. Crop storage was an important subsidiary development, reducing the risk of famine in the months after harvest and in poor harvest years: or rather those risks were reduced once the cat (*Felis domesticus*) had been domesticated to control rats and mice. This may account for the sacred status of the cat in ancient Egypt.[30]

It can be argued that with the cereal technologies in place, and the food supply thus enhanced, other occupations could be undertaken by those no longer needed as food gatherers. But cereal culture produced a necessity for new skills, not just an opportunity. The measurement of land and the weighing of produce for ownership, trading or taxation purposes, led to the need for numbers, weights and measures, and to the means of manipulating numbers and shapes with mathematics and geometry. Agricultural operations are seasonal and timeliness is a requisite for success, as Hesiod noted in the 8th century BC.[50] For peoples living at latitudes away from the equator, or subject to seasonal occurrences of natural events like the flooding of the Nile, the need for agricultural timeliness must have been a powerful incentive for the study of the seasons, astronomy, and the concept of the calendar. In many of the world's religions their festivals, feasts, fasts and holy days became linked to the agricultural year. The name of the cereals comes from that of the Roman corn goddess Ceres, corresponding with Demeter, the Greek goddess of agriculture and corn. The cereals have continued to feature in art and literature down the centuries.[45]

Life is more complicated now than in Neolithic times, and so are modern societies and cultures. However they still depend upon the grains, even though that dependence may now be less obvious, at least in developed countries. The important grain yielding species are: wheat (*Triticum aestivum* and *T. turgidum*, for bread or for pasta respectively), maize (*Zea mais*), barley (*Hordeum sativum*), oats (*Avena sativa*), rice (*Oryza sativa*), sorghum (*Sorghum vulgare*) and millet (*Panicum miliaceum*). World annual production of wheat is about 590 million tonnes (1998 figures from the Food and Agriculture Organisation). Total world production of all the cereals in that year was estimated at about 2 billion tonnes. Cereals now occupy about 690 million hectares of land world wide.[31]

Not all commentators have regarded the revolutionary changes in lifestyle brought about by cereal culture as having been unequivocally beneficial. Settled agriculture has been accused by some writers of providing a less balanced diet than that of the hunter gatherers, who would have had variety in their food. Some cultures became too dependent on one crop, and therefore they became vulnerable to crop failures.[32]

Oats.

Wheat.

Stacking the cereal harvest, 1940. Both pictures from the collections of the Department of Manuscripts and Special Collections, University of Nottingham.

Sheaves of wheat awaiting carting; 1940. From the collections of the Department of Manuscripts and Special Collections, University of Nottingham.

In an ambitious world food history project compiled by Kiple and Ornelas the editors and other contributors pointed out that the settled mode of life, following the development of farming, brought concentrations of people together, with a consequent increase in diseases and in the contamination of water supplies. High carbohydrate diets may have worsened tooth decay problems.[11]

Where populations, or the deprived sections of them, became too dependent on cereals, and did not have enough meat, there were potential dietary shortages of vitamin B_{12} and of readily absorbed iron. The acceptability of the change to cereals may have been assisted by the discovery of salt.[2]

There is an argument that farming became less settled than hunter gathering, rather than more, because as populations grew cultures based on agriculture were forced into a relentless pursuit of new land to cultivate.[34] However few would deny that hunter gathering could only support small populations at low cultural levels. Furthermore if food variety is any criterion of the success of a food provision system, then the modern supermarket must compare rather well with even the most generous view of hunter gathering.

Bread was a significant part of the diet, and of the household budget, for many in England until about the early 20th century. But it was not always bread made with wheat. Until around the 18th century wheaten bread was generally for the better off, while cheaper bread was made with barley, sometimes mixed with pulses. Such bread would have made a useful contribution to protein supply, but it would have yielded less energy for those doing heavy manual labour. In 1789 Gilbert White, in his famous *The Natural History and Antiquities of Selborne*, became quite lyrical about the fact that by then, in the south of England, all classes were eating wheaten bread, which he considered vastly superior to the earlier bread based on barley or beans. He thought that those living in the uplands, and still dependent on the old bread, were liable to itching and skin complaints.[33] There is evidence that barley bread was still the staple of the rural poor in the Midlands four decades later, in the 1830s.[43]

Adding value to cereals on the farm. This water mill at the Farm and Country Centre, Sacrewell, was built in the 19th Century.

The ability of the ruminant animals, or rather of the microbial populations in their rumens, to digest cellulose, has permitted man to inhabit some of the less hospitable regions of the world, including the uplands and dry areas. This is why many cultures measure their wealth in camels, cattle, sheep or goats. Pigs and poultry, the non-ruminant farm animals, increasingly provide protein for the growing populations of the world, and their feeds are based on cereal grains.

Sugar cane (*Saccharum officinarum*) is one of the world's impressive producers of tropical and sub-tropical biomass, growing up to 6 m high. It has probably been used for chewing for thousands of years. Later sugar became important in food preservation and sweetening, and of course in confectionery. The plant source of sugar in temperate climates, sugar beet (*Beta vulgaris*), is not from the family *Gramineae* but from a different botanical family (see below). Its cultivation in the UK is relatively recent, dating from the 1920s.[35]

Grass seeds were not the only plant material originally gathered by the women and children and later planted. A large number of plants have been domesticated to become today's fruits and vegetables.[3] Although the history of horticulture is beyond the present scope no account of food would be complete without some mention of the plants which have become so important in modern eating and shopping. These days even a small supermarket will regularly offer at least five or six varieties of apple (*Malus x domestica*), and sometimes more than a dozen types of onions (*Allium sepa*) and potatoes (*Solanum tuberosum*).

The legumes were originally gathered before they were cultivated. The broad bean (*Vicia faba*) was an early example, perhaps grown as early as 9,000 years ago. The Egyptians had peas (*Pisum sativum*) and vetches (*Vicia sativa*). The soya bean (*Glycine max*), now so ubiquitously used in food manufacture, came from China. Much later, green peas came from Holland and France about AD 1660, and tomatoes (*Lycopersicon esculentum*) were grown in French and Italian gardens by 1750.[3] The potato arrived in Europe from South America in about 1565, but the idea that Sir Walter Raleigh brought it to England is probably a myth.[36]

In England in Tudor and Elizabethan times meat was the food of the better off while vegetables were considered poor man's food. Even the elaborate menus of the 19th century gastronomes relied heavily on meat dishes. Therefore agricultural histories and early texts have little to say about vegetables and fruit, though *Baxter's Library of agriculture* (1846) contained chapters of instruction on the growing of a variety of fruit, vegetables and herbs, and in Roman times Columella documented instruction in their husbandry.[1,41] The Greek invention of grafting was the break through which made orchard scale fruit growing possible.[4] If this is a fair point then it ranks as an invention deserving a mention as one of the keys to the modern food market as we have come to know it.

A clone of the original Bramley apple tree, first planted about 1809 in Southwell, Nottinghamshire. This clone was produced by Nottingham University and is at Merryweather's garden centre in Southwell, where Henry Merryweather had developed the commercial application of the variety in the 19th century. Photographed with permission.

Such was the low status of vegetables that the cultivation of the potato was resisted for two hundred years after its arrival in Europe. Because its edible parts were underground it was seen as devilish and dirty and only fit for peasants. Yet because potatoes required no threshing, milling or baking, they were accessible as staple food to those who lacked capital, whether an entire population, as in Ireland, or amongst those dispossessed of land after enclosure in England. What is more, unlike grain, they provided a certain amount of vitamin C to those unable to obtain any fruits and other vegetables. Larry Zuckerman made the important point that after enclosure small allotment plots of potatoes gave the very poor some degree of independence from the vicissitudes of the price of bread.[36] The potato gradually became a cash crop, and now it would be difficult to imagine a modern diet without it. Chips with everything has become such a staple that we have come full circle, and dietary commentators have started to worry about too much reliance on chips and snacks.

Grape vine growing and wine making are making a come back in England at the time of writing, and a map published in 1998 by the UK Vineyards

Association locates 181 vineyards, mostly in the south, but extending as far north as Yorkshire.[37] In the 17th century vineyards were common in England, however, according to a preface to Lodge's translation of *Palladius on husbondrie*, quoting Barnaby Goodge in 1614.[16] Cobbett frequently commented on the grape crops during his famous rural rides of the 1820s.[44]

Unlike cereals, the potato needs no capital equipment for threshing and milling. It has therefore often been a staple for the poor (Zuckerman, 1998)[36]. When it failed in Ireland there was famine in 1845 to 1847.

Niche products provide consumers with choice and producers with opportunities. An example is wine. Vineyards are reappearing in England. This is the Eglantine vineyard in Nottinghamshire.

No story of the heritage of crop plants, nor for that matter of the agricultural sciences, would be complete without a mention of the seminal work of the Austrian monk, Gregor Mendel, on crossing tall and dwarf peas (*Pisum sativum*). He conducted his experiments from 1856 to 1864 in the garden of the Brno monastery, reporting them to the Brno Natural Science Society in 1865 and publishing a paper in 1866. The rules which he discovered founded the science of genetics, and of practical plant breeding. Sadly his work was ignored for many years, and was not rediscovered until 1900, sixteen years after his death.[38,39]

A fairly short list of botanical families accounts for many of the plant species so far exploited by the human economy. They include the *Gramineae* for the staple cereals and for the grasses upon which the husbandry of ruminant animals depends. The *Leguminosae* provide the nitrogen fixing plants which have been, and still are, pivotal in many crop rotations, and which include protein sources used in human and animal nutrition. From the *Leguminosae* clover (*Trifolium repens*) was a critical contributor to the revolution in crop rotations described in Chapter 1, once fallowing was no longer the means of recovering soil fertility after cereal crops.

The *Cruciferae* account for many of the root and leaf vegetables, including many from one remarkable genus, *Brassica.* The turnip, so important in the history of crop rotations, is *Brassica campestris* and its close relative the swede, important in animal feeding for over a hundred years, is *Brassica napus*. Cobbett called it the Swedish turnip. Oilseed rape is also *Brassica napus*. The genus *Brassica* also includes *Brassica oleracea*, varieties of which include the cabbage, cauliflower, Brussels sprout, kale, kohlrabi and broccoli.

The *Solonaceae* contribute potatoes and tomatoes. Many of the herbs and spices are produced by members of the two families *Labiatae* and *Umbelliferae*, though a notable exception is mustard; another member of that variable and useful genus *Brassica.* The *Umbelliferae* also include the carrot (*Daucus carota*) and the parsnip (*Pastinaca sativa*). The family *Chenopodiaceae* includes that important agricultural species *Beta vulgaris*, which includes sugar beet, fodder beet and beetroot, not to mention the mangold-wurzel of animal feed fame.

Compilations of information on food plants and their histories have been published by Vaughan and Geissler in1997 and by Kiple and Ornelas in 2000.[35,11] Botanical names have been taken from these sources, and from Hubbard for the *Gramineae*.[40] Many of the species mentioned above have interesting histories of domestication of their own, and interesting origins. The carrot, for example, may have, according to Vaughan and Geissler, originated in Afghanistan and spread from there from the 10th and 11th centuries, reaching England in the

15th century.[35] Oddly enough the date of the domestication of the important species *Brassica oleracea* is uncertain.

REFERENCES

1 Baxter, J. (1846) *Baxter's Library of Agriculture.* First published 1830

2 Yudkin, J. (1975) The meat-eating habit in man. In: *Meat.* Edit. Cole, D.J.A. and Lawrie, R.A., Butterworths, London, 3-15

3 Toussaint-Samat, M. (1994) *History of Food.* Translated Bell, A., Blackwell, Oxford

4 Toussaint-Samat, M., Alberny, R., Horman, I. and Montavon, R. (1991) Translated by Jennings, A. *2 Million Years of the Food Industry.* Nestlé S.A., Vevey

5 Diamond, J. (1997) Location, location, location: the first farmers. *Science* **278:** 1243-1244

6 Clutton-Brock, J. (1981) *Domesticated Animals.* Heinemann and British Museum, London

7 Hagedoorn, A.L. (1954) *Animal Breeding.* Crosby Lockwood, London

8 Clutton-Brock, J. (1999) *A Natural History of Domesticated Animals.* Cambridge University Press

9 Ryder, M.L. (1983) *Sheep and Man.* Duckworth, London

10 Gade, D.W. (2000) Cattle. In: *The Cambridge World History of Food.* Edit. Kiple, K.F. and Ornelas, K.C., Cambridge University Press, 489-496

11 Kiple, K.F. and Ornelas, K.C. (2000) *The Cambridge World History of Food.* Cambridge University Press

12 Wing, E.S. (2000) Animals used for food in the past: as seen by their remains excavated from archaeological sites. In: *The Cambridge World History of Food.* Edit. Kiple, K.F. and Ornelas, K.C., Cambridge University Press, 51-57

13 Rixson, D. (2000) *The history of meat trading.* Nottingham University Press

14 Reed, C.A. (1984) The beginnings of animal domestication. In: *Evolution of Domesticated Animals.* Edit. Mason, I.L., Longman, London, 1-6

15 Ryder, M.L. (1984) Sheep. In: *Evolution of Domesticated Animals.* Edit. Mason, I.L., Longman, London, 63-84

16 Lodge, B. (1873) *Palladius on Husbondrie.* Early English Text Society, London

17 Mason, I.L. (1984) Goat. In: *Evolution of Domesticated Animals.* Edit. Mason, I.L., Longman, London, 85-99

18 Epstein, H. and Bichard, M. (1984) Pig. In: *Evolution of Domesticated Animals*. Edit. Mason, I.L., Longman, London, 145-162

19 Wiseman, J. (1986) *A History of the British Pig*. Duckworth, London

20 Wiseman, J. (2000) *The Pig. A British History*. Duckworth, London

21 Malcolmson, R. and Mastoris, S. (1998) *The English Pig. A History*. Hambledon Press, London

22 Epstein, H. and Mason, I.L. (1984) Cattle. In: *Evolution of Domesticated Animals*. Edit.Mason, I.L., Longman, London, 6-27

23 Fraser, A. (1959) *Beef Cattle Husbandry*. Crosby Lockwood, London

24 Troy, C.S., MacHugh, D.E., Bailey, J.F., Magee, D.A., Loftus, R.T., Cunningham, P., Chamberlain, A.T. and Bradley, D.G. (2001) Genetic evidence for Near-Eastern origins of European cattle. *Nature* **410:** 1088-1091

25 Lutwyche, R. (1998) Rabbiting on. *The Ark*, Spring 1998, 23

26 Gloucester Cattle Society (2001) *Tales of Gloucesters. The Rescue of a Cattle Breed*. Gloucester Cattle Society, Oxford

27 Stanley, P. (1995) *Robert Bakewell and the Longhorn Breed of Cattle*. Farming Press, Ipswich

28 Heun, M., Schafer-Pregl, R., Klawan, D., Castagno, R., Accerbi, M., Borghi, B. and Salamani, F. (1997) Site of einkorn wheat domestication identified by DNA finger printing. *Science* **278:** 1312-1314

29 McKenzie, K.S. (1993) Breeding for rice quality. In: *Rice Science and Technology*. Edit. Marshall, W.E. and Wadsworth, J.I., Marcel Dekker Inc., NY, 83-111

30 Wright, M. and Walters, S. (1980) *The Book of the Cat*. Pan Books, London

31 Food and Agriculture Organisation (1999) *Production World Yearbook*, FAO, Rome

32 Crawford, M. and Ghebremeskel, K. (1996) The equation between food production, nutrition and health. In: *Food Ethics*. Edit. Mepham, B., Routledge, London, 64-83

33 White, G. (1789) *The Natural History and Antiquities of Selborne*. 1901 edition, Miall, L.C. and Fowler, W.W., Methuen, London

34 Brody, H. (2001) *The Other Side of Eden: Hunter-gatherers, Farmers and the Shaping of the World*. Faber and Faber, London

35 Vaughan, J.G. and Geissler, C.A. (1999) *The New Oxford Book of Food Plants*. Oxford University Press

36 Zuckerman, L. (1998) *The Potato*. Macmillan, London

37 Robinson, J. (1998) *Vineyards of England and Wales*. UK Vineyards Association, Saxmundham

38 Watson, J.A.S. and West, W.J. (1956) *Watson and More's Agriculture: The Science and Practice of British Farming*. 10th edition. Oliver and Boyd, Edinburgh, 518-524

39 Tyler, C. (2000) *In Mendel's Footnotes*. Cape, London
40 Hubbard, C.E. (1984) Revised by Hubbard, J.C.E., *Grasses,* Penguin, London
41 Columella, L.J.M. Translated into English 1745. *Of Husbandry in Twelve Books and His Book Concerning Trees*. Millar, London
42 Zohary, D. and Hopf, M. (1994) *Domestication of Plants in the Old World*, Clarendon Press, Oxford
43 Arch, J. (1966) *The Autobiography of Joseph Arch*. Edit. O'Leary, J.G., MacGibbon and Kee, London. First published 1898 as *Joseph Arch: The Story of His Life*
44 Cobbett, W. (1830) *Rural Rides*. 1967 edition with introduction by Woodcock, G., Penguin, Harmondsworth
45 Charles, D.R. (2001) *Plants That Changed History*. Keyworth and District Local History Society. Occasional Publication No.8
46 Youatt, W. (1834) *Cattle; Their Breeds, Management and Diseases*. London.
47 Long, J. (1886) *The Book of the Pig*. Upcott Gill, London
48 Department for Environment, Food and Rural Affairs (2001) *Agriculture in the United Kingdom*
49 Close, W.H. and Cole, D.J.A. (2000) *Nutrition of Sows and Boars*. Nottingham University Press
50 Lattimore, R. (1959) *Hesiod*. University of Michigan Press
51 Jellings, A.J. and Fuller, M.P. (1995) Arable cropping. In: *The Agricultural Notebook*. Edit, Soffe, R.J., Blackwell Science Ltd., Oxford, 150-193

3

THE INTENSIVE LIVESTOCK INDUSTRIES

USING THE POULTRY SECTOR AS AN EXAMPLE

Both the pig and the poultry sectors of the British livestock industry developed mainly intensively for several decades. That is to say there was some emphasis on indoor systems. The poultry industry offers a useful and fairly complete example, so this chapter takes the story of the development of the UK poultry industry as its theme, and as an example of the history of an intensive sector of British agriculture.

Despite the suitability of the story of poultry to the purpose of an account of the intensive sectors, it may seem strange to devote a separate whole chapter to its history, but there are three more reasons for its inclusion. Firstly it happens to be a sector whose history I know a bit about. Secondly it provides a particularly interesting example of an agricultural sector coping with changing times. It realised early in its development that it is part of the food supply chain, operating in response to consumers. Thirdly there are some myths to dispel.

Outdoor pigs. Photo: DJA Cole.

Indoor pigs. Photo: J Gadd.

It is an important national food source, a major provider of rural employment, and a customer of the arable industries. Therefore a review article on the subject, mainly covering recent history, was published on the occasion of the 50th anniversary of the UK Branch of the World's Poultry Science Association.[1,2,3] The remarks which follow explore the story a little further.

GENETIC HISTORY

Chickens

The genus *Gallus* probably dates from about 8 million years ago.[4] This is recent by the standards of the avian order *Galliformes*, which goes back about 35 to 40 million years and includes chickens, turkeys, pheasants, grouse, partridge, quail, guinea fowl and ptarmigan.

The domestic fowl (*Gallus domesticus*, or, if regarded as a sub-species, *Gallus gallus domesticus*) may be derived from more than one species, which probably helps to account for the rich variation available to breeders, and the range of attributes observed in this remarkably versatile bird. The enormous difference between a laying hen and a broiler springs to mind, not forgetting the vast differences in plumage, body size and appearance amongst the traditional breeds. There are bantams with mature weights of less than ½ kg, and at the other end of the scale there is the Jersey Giant with cocks weighing in at nearly 6 kg.[5,36,72]

Species thought to be possible ancestors of the modern chicken are *Gallus gallus, G. lafayettei, G. sonnerati* and *G. varius.* The red jungle fowl, *Gallus gallus*, appears to be the main progenitor, though the Asiatic and the Mediterranean breeds may have different origins.[4] There is a view that the jungle fowl was not the only progenitor. The heavy breeds may have originated from Chinese fowls such as the Cochin and the Brahma.[6]

A modern jungle fowl; probably resembling the ancestors of the chicken. Photo: J.G. Corder, World Pheasant Association.

Hybrids would probably have occurred between the early domesticated and wild chickens, and this helps to account for the variation between breeds (Hagedoorn, 1954).[7] On the other hand some authorities do not accept a multi-species origin, on genetic and archaeological grounds.[8] Genetic affinities with the pheasants may be important.[9]

All of the probable progenitors are from south east Asia, in the region that is now Thailand. The natural range of the jungle fowl is bounded by the 10°C January temperature line. The wild red jungle fowl is indigenous to the Himalayan foothills and to much of northern India.[9,10] So perhaps we should not have been so surprised when it was confirmed in the 1960s and 1970s that modern chickens require a relatively high temperature.[11,12] Notable 19th century writers had been aware of the dangers of cold weather and warned us of it colourfully. The Hon. Mrs Arbuthnot (1871) advised: "In cold weather, feed liberally on toast soaked in ale. Fowls are by no means abstainers, but heartily enjoy their beer, nay, even wine, when suffering from debility".[13] Centuries before that the Roman author Columella had written: "Hen-houses ought to be placed in that part of the manor-house which looks to the sun rising in the winter; let them be contiguous either to the oven, or to the kitchen."[14]

Based on recent archaeological evidence domestication is thought to have first occurred in south east Asia by 6,000 BC.[10] This is much earlier than had been believed. The chicken was probably taken north and established in China, reaching Europe by uncertain routes, but perhaps *via* Russia. Its arrival in Britain was pre-Roman, and Julius Caesar mentioned that it was already present.[15] The Dorking breed was in England in Roman times. Its introduction may have been *via* the Phoenicians. However the Romans almost certainly also brought their own chickens with them. Chickens reached the Americas either with the Spanish conquest or much earlier by way of the Pacific.[4]

Early breeding after domestication was probably for traits of little relevance today, and may have included selection for cock crow! Roman and Zoroastrian literature shows that the ancients were much impressed by this attribute.[16] The chicken was used as a sacrificial animal, so presumably its appearance may have mattered, and its spread was probably more for cock fighting purposes than for food. It is worth remembering that some of the characteristics in these early selection programmes, such as aggression for fighting, are inconsistent with current husbandry requirements.

Analysis of ovoglobulin (an egg white protein) suggests that the Asiatic breeds evolved separately from the European.[4]

Turkeys

The wild turkey, *Meliagridis gallopavo*, is indigenous to the new world and was probably domesticated in Mexico (Appleby *et al*., 1992).[17] It was brought to Europe at the time of the Spanish conquest of South America, though Columbus may have taken specimens back to Spain a little ealier in 1498.[18] It reached Germany in 1530 and England by 1541.[19] There is some confusion over its arrival in Europe, however, and even its name adds to the confusion. There has never been any suggestion that Turkey was the original home of the bird. The story is that merchants returning to England from Turkey were offered the meat in Cadiz on the way home, though Seville may have been the stopover.[20,21] The merchants chose the name having been understandably daunted from attempting to pronounce the Mexican name *uexolotl*. Even if this story is not the whole truth it is at least colourful.

In Stuart times large flocks of both turkeys and geese were driven on foot from East Anglia to London, setting off in August and arriving at the markets at Smithfield and Leadenhall in time for Christmas.[22] Their feet had to be protected with tar. East Anglia became associated with turkey and goose production because of the availability of stubbles after the harvest in a cereal growing region.[23]

The Norfolk Black has been regarded as a breed similar to the original turkeys introduced into England. It has therefore been listed, along with the Cambridge Bronze, in a directory of traditional foods.[5,23]

Modern turkeys. Photo: ADAS.

Ducks

All common domesticated ducks were probably descended from the wild mallard, *Anas platyrhynchos,* and were domesticated in south east Asia or China. Domestication may not have occurred in Europe before the middle ages despite references to domestication in China at least 3,000 years ago.[17,24,25]

The Muscovy duck is a different species, and is derived from *Cairina moschata*, a perching duck from South America.[25,26]

Geese

Geese of the genus *Anser* were probably domesticated between 3,000 and

2,700 BC according to Romanov.[27] The greylag goose, *Anser anser*, is probably the ancestor of European domestic geese, though the Asiatic swan goose, *Anser cygnoides,* may also have been involved in some domestications. The Egyptians crammed geese to obtain fat livers as early as 2,700 BC.[27,28]

Goose is a traditional English dish with a very long heritage. It was eaten at Michaelmas (29 September) and at Christmas. The Michaelmas goose may be a perpetuation of very old Celtic feasts as thank offerings for the harvest. There were autumn goose fairs, the most famous being at Nottingham, originating in the 13th century and still going strong today, though nowadays it is a fun fair.[23] Until the arrival of the turkey, goose was the big roast for special occasions. In the 19th century and before few people had a big enough oven at home and the goose was often taken to the baker's for cooking.[21]

Pigeons

Although pigeons are not farmed in Britain nowadays, they were once an important winter meat source. Dovecotes housing rock doves, or blue rock pigeons, *Columba livia*, were common accessories of manor houses. The young, or squabs, were considered a delicacy.[29] Eighteenth century agriculturalists valued pigeon manure. Ellis's husbandry, published in 1772, declared that "Pigeons-dung is indisputably the hottest dung for all wet clays and moist loams, where, in a peculiar manner, the fiery salts act most potently in bringing forward the growth of all grains, grass and trees."[30]

THE FOUNDATIONS OF AN INDUSTRY

The selection and development of the chicken as a food animal probably originated in the Greek and Roman civilisations; and the Romans had a well structured poultry industry, complete with proper husbandry systems and technical authors.[16] Performance levels were said to be quite good; whatever that might mean. Poultry had arrived in Britain by Roman times, as indicated by bones found in London.[31] The best known poultry author of the time was Columella, translated into English in 1745, who presaged modern hybrids when he wrote in his *Book the Eighth*, "the bastard chickens are the best, which our own Italian chickens have brought forth, having conceived them by foreign males".[14] Three centuries after Columella, another Roman named Palladius, in the 4th century, wrote that "Two cruses of half-boiled barley is one day's food for a hen at large". Presumably "at large" was an early technical term for free range.[32]

After the decline of Rome the organised poultry industry also declined and the birds assumed the role of scavengers. In medieval times breed selection for cock fighting was still taking place. In medieval London eggs were on sale at 1d for 10, (0.4p); a capon baked in a pasty at 8d, roast plover 2½d, heron 18d, 10 finches 1d, 3 thrushes 2d, and 5 larks 1½d.[33] Medieval poultry products certainly offered consumers variety. The trade was significant enough for the Poulterers Livery Company to have been established in 1124. In 1393 monks established swan farming at Abbotsbury, Dorset, and a swannery can still be visited there.

In the 13th century manor the poultry yard was under the care of the dairywoman. The text book *Hosebonderie* suggested that each hen was to answer for 115 eggs and 7 chickens.[34] The Staffordshire chronicler, Plot, was informed by Thos. Broughton Esq. in 1686 that a certain Ann Biddulph of Edgiall had a hen that would lay 3 eggs a day.[35]

Not until the 19th century did poultry become a specialised enterprise again and the textbooks still quoted Columella.[16] However breed societies and poultry shows appeared from the mid 19th century. Interestingly the forerunner of the present day trade event (Poultry Fair), was still called the Poultry Show until the 1960s.

At the time of the great ages of livestock improvement of the 18th and parts of the 19th centuries poultry missed the attention of the likes of Robert Bakewell of Loughborough (1725-1795). Yet Bakewell's method of crossing inbred lines of cattle and sheep, in which he had fixed desirable characteristics, was similar to the methods used by international breeding companies to produce hybrid poultry two hundred years later.

Dorking, Surrey, shares its name with a chicken breed.

The development of the traditional poultry breeds was a succession of introductions and re-introductions of foreign stock. There was the Leghorn from Italy and USA, the Cochin from China, several breeds of Mediterranean origin, and of course the Rhode Island Red.[36] Meanwhile the Dorking, the Sussex, the Derbyshire Red Cap and many more British breeds flourished. Much of the interest of the breeders until the 19th century was in exhibition rather than in production. Large numbers of breeds are still extant. The 5th edition of *British Poultry Standards* lists 93 breeds of chicken.[5]

THE BEGINNINGS OF THE MODERN POULTRY INDUSTRY

Three accounts of the development of the modern poultry industry have been published recently, and some of this chapter is based on them.[37,38,39] Archive material and artefacts illustrating the last 100 years are held in the poultry museum collection at the Farm and Country Centre, Sacrewell, Peterborough.[2,3]

Commercial poultry farming was encouraged after World War I, when egg production was regarded as a suitable occupation for returning troops.[37] Several counties had schemes for 3 acres and a cow, and sometimes 100 hens were seen as an alternative. Every county had its laying trials by the 1930s, even relatively urban counties like Middlesex. Local authorities fielded poultry instructors and published poultry textbooks and manuals.[40]

Every county also had its commercial pedigree breeders, some of whom became quite sizeable businesses. Many survived until the hybrids arrived in the 1950s and 1960s. The Scientific Poultry Breeders Association kept a register, and copies for 1936 and 1937 are preserved in the museum archives. A few of the traditional breeders, as they are now called, are still trading at the time of writing. Some sell equipment and housing as well as birds. The Rare Breeds Survival Trust has registers of breed availability. By the late 1990s interest in niche markets had given some of the traditional breeds new opportunities and markets, and their technology was once again the subject of research projects.[41]

The poultry industry was so important during the inter-war years that when the fourth world congress of the World's Poultry Science Association met at the Crystal Palace in 1930 it was opened by the Duke of York and nearly 100,000 visitors attended.[38]

During World War II poultry played an important role converting waste and scraps into valuable food, thus saving imports and scarce shipping. There were 11.5 million hens in gardens and on allotments.[42] Numerous books and pamphlets offered instruction to both commercial and domestic poultry keepers. Poultry feed was rationed, due to shortage of grain, and remained so until 1954. It was only after this that the expansion of businesses and of production became possible.

Pigs and poultry and the war effort. This World War II swill mixer was used for making scraps into feed. Photographed at Sacrewell Farm and Country Centre.

Consumers often perceive battery cages as a very modern idea, but they were used, usually by smallholders, in small but increasing numbers from their introduction into Britain in about 1925. Sometimes their use was to facilitate individual bird recording, but it was also to solve problems such as minimising coccidiosis and fowl sickness of the land, minimising floor eggs and dirty eggs, making depletion easier, reducing capital and labour cost and reducing problems caused by the birds pecking each other. There are records of battery cages in use in the USA even earlier than 1925.[43] A Professor Halpin of Wisconsin used them in 1911, though they were not widely commercially manufactured until about 1930 in the USA or in Britain. Early attempts in Britain were not always successful, including a test in 1929 by the Poultry Association of Great Britain.[44] Thus during much of the early growth of the commercial British poultry industry most of the birds were on range or in semi-intensive systems such as straw yards and pebble yards. The technology of the period was documented in 1948 in a famous textbook by Leonard Robinson.[45] Free range of the period often involved arks moved daily, and required a great deal of labour and land. Many of the problems which battery cages were introduced to solve re-emerged as challenges during modern attempts to develop commercial alternatives to cages.

The search for alternatives to cages, in response to consumer wishes, caused the revival of some old established principles, but with the advantages

A small incubator of about the 1940s, and now in a collection of artefacts at Sacrewell Farm and Country Centre.

Not all the old fashioned ways would have pleased modern consumers. Cramming machines like this were used for forced feeding table poultry until the late 1940s or early 1950s. Table cockerels were sometimes grown in cages at the time, though modern broilers never are. Both photos at Sacrewell Farm and Country Centre.

of the application of modern knowledge of poultry biology and behaviour. For example, in several European countries there has been much interest lately in enriched cages. Various versions of them have been tried. Some have nesting areas and perches, and sometimes dust baths. The 1990s British versions were based on 1970s experience with prototypes called aviaries, which had platforms within litter houses, and with the "get-away cage", which had a nest box and perches, and feeders and drinkers at two levels. But these

Harvesting eggs for the war effort, 1939. Photo: G.S, McCann, Uttoxeter. From the collections of the Department of Manuscripts and Special Collections, University of Nottingham.

systems are not new. A system very like the current enriched cages, was built by a Captain Gregory in Sussex in 1929. It had 10 birds in a square cage with a nest box attached. By 1932 there were 600 birds in such cages, and it was successfully copied in 1948. The limiting factors were broodiness and eggs not laid in the nest box.[43]

The Women's Land Army on poultry work, 1939. Photo: G.S. McCann, Uttoxeter.
From the collections of the Department of Manuscripts and Special Collections, University of Nottingham.

A free range ark of the type used in the 1950s and before and still used by smallholders. Modern commercial free range requires accommodation for larger numbers of birds. Photographed at Sacrewell Farm and Country Centre.

Range accommodation in the 1940s. Photo: ADAS.

Consumers may, perhaps, regard large poultry flocks as a modern idea, but history tells a different story. Large dovecotes were common attachments to stately homes. A 16th century dovecote at Willington, Bedford, housed 1500 pigeons to provide winter meat.[22] The great dovecote at the Grantchester Estate, Cambridge, provided 2,000 to 3,000 pigeons a year.[29]

Range arks circa early 1950s. Photo: ADAS.

THE MODERN INDUSTRY: THE CONTRIBUTION OF SCIENCE

The development of the modern industry, with its large businesses and its affordable products for consumers, could be said to have begun in 1954, when expansion became possible after the ending of poultry feed rationing.[46] Throughout the period since 1954 poultry science played a facilitating role. Some of the scientific contributions considerably predated 1954 however, and the following is a summary of some technical landmarks. (Note that in a recent book from this publisher the period before 1954 was defined as the *traditional period* of the British poultry industry, and after 1954 as the *conventional period.*)[41]

Serious attempts at breeding for performance instead of for appearance were made in the 1890s. By the 1920s the agricultural colleges were improving rations and the commercial breeders were getting together to fund research into several disciplines, including nutrition and breeding. Copies of *"Eggs"*, the journal of the Scientific Poultry Breeders Association, make impressive reading. By the 1920s and 1930s its authors already understood the elements of poultry nutrition as we would recognise them today.[58]

In northern latitudes, including Britain, eggs used to be a seasonal food, difficult for consumers to obtain in winter. It was not warm accommodation but supplementary electric lighting which was the key to taking the seasonality out of the supply of eggs to UK consumers. Before the advent of supplementary lighting large tonnages of eggs were imported during the winter, and as recently as the 1950s eggs were preserved in pots of water-glass (sodium silicates) for winter use. Older preservation methods included storage in ice houses and painting with oil or resin.[22]

Supplementary lighting was investigated in detail as early as 1925, when Sir Edward Brown reported on the first experiments at Reading University on lighting for winter eggs, though this was not the first attempt. There were reports of attempts by Spanish farmers 100 years earlier.[47,48] In the USA a publication called the *Reliable Poultry Journal*, of Illinois, printed an article in 1915 by E.C.Waldorf, MD, who had used gas burners controlled by an automatic timer for the purpose in 1889.[73] In the late 1920s winter lighting programmes were developed at Harper Adams College.[49,50]

Some of the early writers reviewed by Sir John Hammond in 1960, in a book originally published in 1940, thought that the effect of lengthening days was simply to give the birds longer to eat.[51,52] But by the 1930s it had been realised that the seasonality of egg production was due to the seasonal changes in daylength and therefore hormonal effects.[53]

Thus by the 1930s it was known that lights could be used to simulate the natural increase in daylength of springtime.[53] In the late 1950s and early 1960s at Reading University, and in the Alabama Agricultural Experimental Station, practical lighting programmes were developed, and these were soon almost universally applied by the UK poultry industry.[54,55]

A modern poultry house at ADAS Gleadthorpe, with insulated walls and roof and electric fan ventilation, i.e. the so-called controlled environment system of production. Photo: ADAS.

There is a collection of books of the early days of the scientific era in the archives of the poultry museum. They are a curious mixture of the scientific and the anecdotal approaches, but the total number of publications is impressive. The collection contains 64 titles published from 1920 to 1950 inclusive, and no doubt there were many more which have escaped preservation. Many companies in the ancillary supply trades produced guides and manuals. This outpouring of books is even more impressive if allowance is made for the fact that from about 1939 to 1953 there was a paper shortage. The *magnus opus* of the egg science of those days was a book compiled by Romanov and Romanov, published in 1949, running to 918 pages, and still a valuable source of scientific information today.[56]

By the 1950s most of the important vitamins had been identified and understood, and mystique in ration formulation had been replaced by the concept of meeting nutrient requirement at least cost. Least cost feed formulation must have been one of the earliest serious industrial applications of computer technology, having been a commercial proposition by the early 1960s.

The year 1956 saw the introduction of broiler hybrids from the USA. The late 1950s and early 1960s were times of rapid development of the international corporate poultry industry, of integration, and of the use of hybrids. Books such as *"Poultry - a modern agribusiness"* written by Geoffrey Sykes in1963, after studies in America, were characteristic of the pride which the industry took at the time in leading agriculture into a more business based era.[57] The scientific journal *British Poultry Science* was launched in 1960 under the guidance of Alec King of the National Agricultural Advisory Service (NAAS).

The shift to indoor systems accelerated during the 1960s, while holding size increased and the numbers of holdings decreased. As time went on it became necessary for holdings to become larger and fewer in order to be commercially viable in the commodity egg and chicken trades. This was because product prices steadily fell in real terms. Dr. Rupert Coles, MAFF's Chief Poultry Officer, writing advisory material in 1966, noted a decline from over 250,000 holdings with poultry in 1957 to less than 160,000 in 1965. By 1965 the number of laying flocks of over 5,000 birds was still only about 2%, but these flocks already contained 30% of the birds.[55] MAFF reported that in the UK by 1999 79% of layers were in flocks of over 20,000 birds, though there were still 2.2 million birds in flocks of under 5,000 (see Table 3.1). In the broiler sector large flocks also became the norm.[59] In the future intensive pig and poultry unit sizes will be affected by the Integrated Pollution Prevention and Control (IPPC) regulations and by directives requiring Best Available Techniques (BAT).[79]

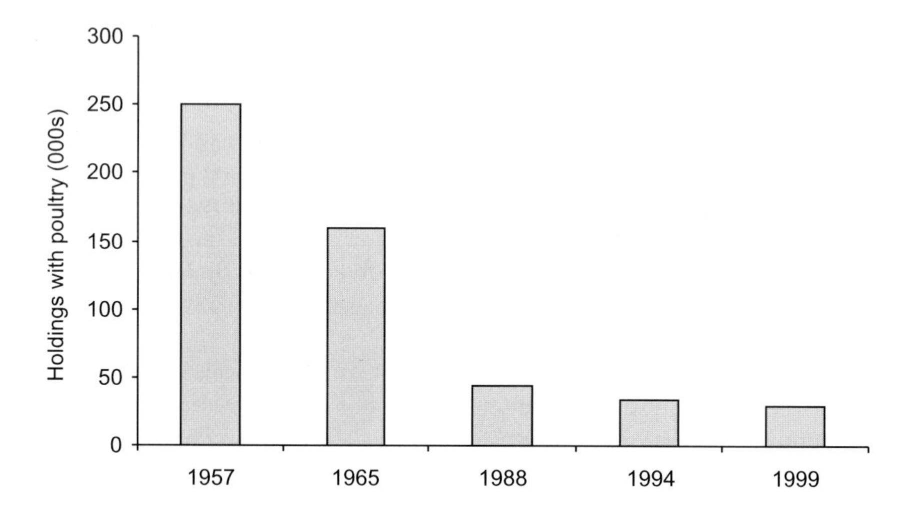

Figure 3.1 Declining numbers of businesses. Holdings with poultry, thousands. Data based on a table of Gordon and Charles (2002)[41] and reproduced with permission.

TABLE 3.1 INCREASING BUSINESS SIZE. % OF NATIONAL LAYING FLOCK IN FLOCKS ABOVE THRESHOLD SIZES

	% in flocks over 500 birds[71]
1948	11
1955	22
	% in flocks over 20,000[59,67]
1988	70
1994	78
1999	79

As the numbers of birds kept per house began to increase beyond about 50 or so it was soon realised that moisture outputs from excretion and respiration were considerable. This moisture could easily lead to condensation on the walls and roof and to poor litter, unless ventilation and insulation were deliberately provided in a systematic way. As early as 1926 Dann in the USA had published figures for the moisture output of the birds and in 1951 C.N. Davies in Britain published physiological calculations of recommended ventilation rates. A century earlier Trotter had noted in 1851 that "...it is therefore of the utmost importance that the ventilation be of the most perfect description".[60,61,62]

By 1955 standard texts, such as the MAFF bulletin on poultry housing, were already recommending insulation for the walls and roof in order to prevent condensation.[63] By the 1960s ventilation with electric fans and thermostats was common, to achieve the birds' temperature requirements more precisely. Thus the so-called controlled environment system of housing had evolved.

The energy crisis of 1974 focussed attention on feed conversion, giving another stimulus to the receptiveness of the industry to knowledge of the biology of the birds and of the physics and engineering of keeping them warm enough for their needs. We have seen in Chapter 1 how the early agricultural scientists began to understand the relationships between feed energy and the climatic temperature needs of animals. The chicken, with its jungle origins, is an extreme example.

Poultry pictures as art. Drawing by Harrison Weir, 1902.[80]

"SUNNING."

LOST!

More pictures from *Our Poultry*. Drawings by Harrison Weir, 1902.[80]

SALES AND MARKETING

As Holroyd pointed out in his Temperton Report of 2001, in the years after World War I poultry production was seen primarily as a provider of employment. After World War II it was regarded as a provider of food. He observed that at the time he was writing in 2001 it was seen as a follower of consumer requirements. These changes have affected its relationship with its markets.[64]

Since their introduction broilers have always been sold on a free market, led by the dictates and requirements of consumers. The egg market was managed during the early development of the modern industry, but later became free. From 1957 to 1967 the British Egg Marketing Board handled egg sales, followed by the Eggs Authority, with limited powers, from 1967 to 1985. In both the eggs and poultry meat sectors there are trade associations providing industry wide promotional and representational work for their members, but they have no statutory powers. In 2001 several of them were combined to form the British Poultry Council. In 1991 the research organisations got together to form the UK Poultry Research Liaison Group to improve communications between the users, the fund providers and the contractors of poultry research.

By the 1980s and 1990s consumers began to influence production methods and there was a revival of interest in free range and barn production (see Table 3.2). Interest in animal welfare was older, probably dating from the publication of *"Animal machines"* by Ruth Harrison in 1964, leading to the Brambell Report of 1965.[65,66]

Many factors have influenced the consumption and sales of poultry products in recent years. The decline in the popularity of the cooked breakfast at home, and of home baking, led to falling sales of eggs in shell from about the mid 1960s till the early 1990s. Total egg sales then stabilised as egg products became more important in food processing and manufacture. Sales of poultry meat increased steadily from the 1950s until recent years, since it is low in fat, cheap and convenient, and lends itself well to modern cooking requirements and to the needs of the added value prepared meals, freezing, catering and fast food trades.

TABLE 3.2 CHANGES IN EGG PRODUCTION SYSTEMS, % OF EGGS

	Cages	*Litter or barn*	*Free range*
1951	8	12	80
1963	27	56	17
1980	95	4	1
1990	85	3	12
1995	86	3	11
1999	79	5	16
2000	70	6	23

Adapted from data of MAFF. Also based on a table by Gordon and Charles (2002) with permission.[41,59,67]

Occasional food scares, such as the *Salmonella* incident of 1989/90, caused temporary perturbations to the general trends. The trade in poultry products is now global and so are many of the companies involved. There is also an international trade in technology and the means of production.

There is now an increasing interest in labelling in order to distinguish British products, because of the efforts made by the industry to meet the ambitions of consumers for welfare, food safety and environmental criteria. Schemes such as Lion brand eggs and the RSPCA Freedom Food brand are examples. Such ventures seem to be successful for products sold as recognisable entities, such as eggs in shell and whole chickens or portions. But once products lose their identity, by becoming ingredients and components of foods, consumers are less aware of the country of origin.

At the time of writing there are expanding niche markets, including for organic produce, and there are organic eggs and poultry meat, though as yet accounting for only a small percentage of the total. Extensively produced table chickens are beginning to appear in test work and on the market. Other niches include corn (*i.e.* maize) fed chickens, barn chickens, several feed based speciality eggs (including some with nutritional enhancements such as omega-3 fatty acids), traditional farm turkeys and many more.

Consumer interest in free range systems led to a switch in research emphasis. This study at ADAS Gleadthorpe in 2000 was on breeds for extensive chicken growing.

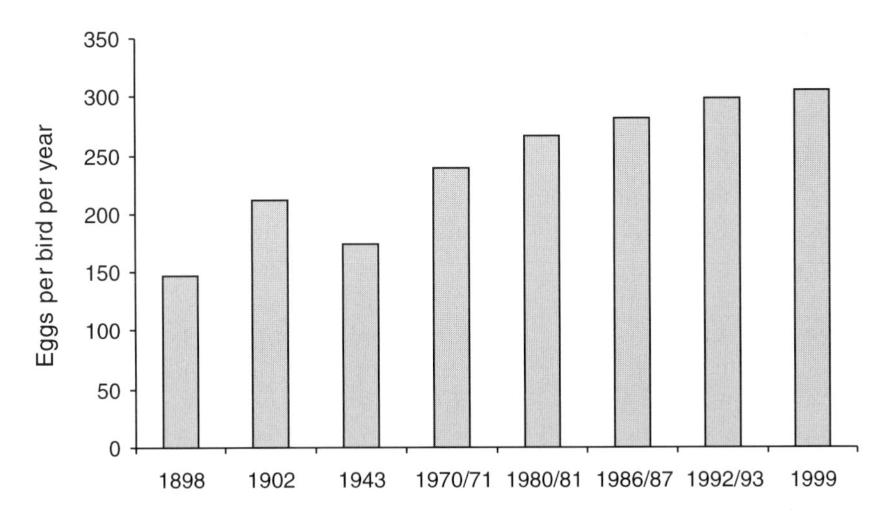

Data of Whittle[68]; except 1=Hawk, pen averages and best of pen in Utility Poultry Club competitions,[40]; 2=Thompson[69]; 3=ADAS data

Figure 3.2 Typical egg production per bird, to 72 weeks of age (numbers of eggs per bird):

TABLE 3.3 TYPICAL WEIGHTS OF BROILERS, AVERAGES OF MALES AND FEMALES

	Liveweight, kg	*Age, days*
1958	1.9	70
1980	1.9	44
1996	2.2	42

Data from ADAS and the poultry museum collection, Farm and Country Centre, Sacrewell

TABLE 3.4 AN EXPANDING INDUSTRY. POULTRY MEAT PRODUCTION IN UK

	000 tonnes carcass weight	*Source*
Pre-war	81	Richardson[70]
1946/46	56	ditto
1953/54	87	ditto
1965/66	412	ditto
1973/74	677	ditto
1989	994	MAFF[67]
1992	1077	ditto
1999	1523	MAFF[59]

Data from ADAS and the poultry museum collection, Farm and Country Centre, Sacrewell

Despite the trends shown in Tables 3.3 and 3.4 official statistics suggest that in the UK in 1999 there were still as many as 25,600 holdings with less than 5,000 laying hens each. The total numbers of birds on these holdings suggests that many of them must have been smallholdings containing very small numbers of birds.[59]

This suggests that poultry may have roles in both the large scale commercial food industry and in small enterprises in the countryside.

PIGS

The development of the pig industry in Britain has followed fairly similar patterns to that of the poultry sector. There was a gradual increase in herd size while the numbers of holdings with pigs tended to fall over recent decades. The sector has been supported by scientifically based breeding and nutrition, and by the matching of indoor environments to the climatic needs of the animals. Producers have increasingly taken account of the needs and preferences of consumers and the public: preferences sometimes leading to legislation. As for poultry the issues have included animal welfare, environmental concerns and product quality. Markets have been affected by demographic and sociological changes and marketing has become a critical discipline. There has been recent interest in outdoor production and in niches such as organic production.

THE BERKSHIRE.

The 19th century breeders developed very fat pigs. From Long's *The Book of the Pig* (1886).[78]

The Black Dorsetshire pig, from Long's *The Book of the Pig*.[78]

Modern pigs, with much more lean and less fat than their forebears. Photo: J Gadd.

The history of the British pig has recently been documented by Wiseman[74,75], and by Malcolmson and Mastoris.[76] There have also been occasional publications on the fascinating histories of some pig products, such as a history of the Melton Mowbray pork pie by Hickman.[77]

REFERENCES

1 Charles, D.R. (1996a) *The Poultry Industry, The Environment and the Rural Economy*. 5th Temperton Fellowship Report. Harper Adams Agricultural College

2 Charles, D.R. (1996b) The Museum of the British Poultry Industry. *Journal of the Royal Agricultural Society of England* **157:** 197

3 Charles, D.R. (1996c) A brief history of the UK poultry industry. In: *United Kingdom Branch 50th Anniversary*. Edit. Fisher, C. and Hann, C.M., World's Poultry Science Association, 45-51

4 Stevens, L. (1991) *Genetics and Evolution of the Domestic Fowl*. Cambridge University Press

5 Roberts, V. (1997) *British Poultry Standards*. 5th edition. Blackwell Science Ltd., Oxford

6 Bland, D.C. (2002) Letter to *The Ark* **30:** (1) 32

7 Hagedoorn, A.L. (1954) *Animal Breeding*. Crosby Lockwood, London

8 Blench, R. and MacDonald, K.C. (2000) Chickens. In: *The Cambridge World History of Food*. Edit. Kiple, K.F. and Ornelas,K.C., Cambridge University Press, 496-498

9 Coote, J. (2001) Which came first... *The Ark* **29:** 134-135

10 West, B. and Zhou, B-X. (1989) Did chickens go north? New evidence for domestication. *World's Poultry Science Journal* **45:** 205-218

11 Payne, C.G. (1966) Environmental temperature and egg production. In: *The Physiology of the Domestic Fowl*. Edit. Horton-Smith, C. and Amoroso, E.C., Oliver and Boyd, Edinburgh, 235-241

12 Emmans, G.C. and Charles, D.R. (1977) Climatic environment and poultry feeding in practice. In: *Nutrition and the Climatic Environment*. Edit. Haresign, W., Swan, H. and Lewis, D., Butterworths, London, 31-50

13 Arbuthnot, Mrs. (1871) *The Henwife: Her Own Experience in Her Own Poultry-yard*. 9th edition. T.C. Jack, Edinburgh

14 Columella, L.J.M., translated into English 1745, *Of Husbandry in Twelve Books and His Book Concerning Trees*. Millar, London

15 Crawford, R.D. (1984a) Domestic fowl. In: *The Evolution of Domestic Animals*. Edit. Mason, I.L., Longman, London, 298-311

16 Wood-Gush, D.G.M. (1959) A history of the domestic chicken from antiquity to the 19th century. *Poultry Science* **38:** 321-326

17 Appleby, M.C., Hughes, B.O. and Elson, H.A. (1992) *Poultry Production Systems. Behaviour, Management and Welfare*. CAB International, Wallingford, 3-22
18 Crawford, R.D. (1984b) Turkey. In: *The Evolution of Domesticated Animals*. Edit. Mason, I.L., Longman, London, 325-334
19 Brant, A.W. (1999) A brief history of the turkey. *World's Poultry Science Journal* **54:** 365-373
20 Tannahill, R. (1988) *Food in History*. Penguin, London
21 Toussaint-Samat, M. (1994) Translated by Bell, A. *A History of Food*. Blackwell, Oxford
22 Paston-Willimas, S. (1993) *The Art of Dining. A History of Cooking and Eating*. National Trust Enterprises Ltd., London
23 Mason, L. and Brown, C. (1999) *Traditional Foods of Britain. An Inventory*. Euroterroirs. Prospect Books, Totnes
24 Clayton, G.A. (1984a) Common duck. In: *The Evolution of Domesticated Animals*. Edit. Mason, I.L., Longman, London, 334-339
25 Ashton, C. and Ashton, M. (2001) *The Domestic Duck*. Crowood Press, Marlborough
26 Clayton, G.A. (1984b) Muscovy duck. In: *The Evolution of Domesticated Animals*. Edit. Mason, I.L., Longman, London, 340-344
27 Romanov, M.N. (1999) Goose production efficiency as influenced by genotype, nutrition and production systems. *World's Poultry Science Journal* **55:** 281-294
28 Ashton, C. (1999) *Domestic Geese*. Crowood Press, Marlborough
29 Rixson, D. (2000) *The History of Meat Trading*. Nottingham University Press
30 Ellis, W. (1772) *Ellis's Husbandry*. Nicoll, London
31 Sykes, A.H. (1992) Bone idle. *British Poultry Science* **33:** 1123-1124
32 Lodge, B. (1873) *Palladius on Husbondrie*. Early English Text Society, London
33 Davis, D. (1966) *A History of Shopping*. Routledge and Kegan Paul Ltd., London
34 Prothero, R.E. (1936) *English Farming Past and Present*. Heinemann and Frank Cass & Co., London
35 Plot, R. (1686), republished in facsimile by E.J. Morten, Manchester (1973). *The Natural History of Staffordshire*, 235
36 Anon (1921) *The Book of Poultry*. Routledge, London
37 Hewson, P.F.S. (1986) Origin and development of the British poultry industry: the first hundred years. *British Poultry Science* **27:** 525-540
38 Telford, M.E., Holroyd, P.H., and Wells, R.G. (1986) *History of the National Institute of Poultry Husbandry*. Nuffield Press, Oxford

39 Whittle, T.E. (1998) *A Triumph of Science: A 70 Year History of the UK Poultry Industry*. Poultry World
40 Hawk, W. (1910) *Poultry Keeping for Profit*. Cornwall County Council
41 Gordon, S.H. and Charles, D.R. (2002) *Niche and Organic Chicken Products: Their Technology and Scientific Principles*. Nottingham University Press
42 Hardyment, C. (1995) *Slice of life. The British Way of Eating Since 1945*. BBC Books
43 Blount, W.P. (1951) *Hen Batteries*. Ballière, Tindall and Cox, London
44 Robinson, L. (1950) *Battery Egg Production*. S.P.B.A. Supplies Ltd., Woking
45 Robinson, L. (1948) *Modern Poulty Husbandry*. Crosby Lockwood and Son Ltd., London
46 Charles, D.R. and Tucker, S.A. (1997) The poultry industry in the United Kingdom. *Journal of the Royal Agricultural Society of England* **158:** 175-183
47 Hutchinson, J.C.D. (1956) Control of seasonal variation in egg production of hens. *Nature* **177:** 795-796
48 Morris, T.R. (1999) Sexual maturity, lighting and layer performance. *Poultry Lighting Seminar*, University of Bristol, 10-12
49 Parkhurst, R.T. (1928) Artificial light for late hatched pullets. *Eggs*, Scientific Poultry Breeders Association, December 1928, 270-271
50 Rhys, I.W. and Parkhurst, R.T. (1931) National Institute of Poultry Husbandry Harper Adams, Bulletin No. 6
51 Hammond, J. (1960) *Farm Animals*. Edward Arnold, London
52 Fairbanks, F.L. and Rice, J.E. (1924) *Artificial Illumination of Poultry Houses for Winter Egg Production*. Cornell Extension Bulletin 90
53 Whetham, E.O. (1933) Factors modifying egg production with special reference to seasonal changes. *Journal of Agricultural Science* **23:** 383-419
54 Morris, T.R. and Fox, S. (1958) Light and sexual maturity in the domestic fowl. *Nature* **181:** 1453-1454
55 Coles, R. (1966) Size changes in laying flocks. *The Poultry Review*. May and Baker Ltd., Dagenham, **6**
56 Romanov, A.L. and Romanov, A.J. (1949) *The Avian Egg*. Wiley, New York
57 Sykes, G. (1963) *Poultry - a Modern Agribusiness*. Crosby Lockwood and Son Ltd., London
58 Meech, R. (1912) *Eggs*. Official Organ of the Scientific Poultry Breeders' Association
59 Ministry of Agriculture, Fisheries and Food (2000) *Agriculture in the United Kingdom 1999*. Stationery Office, London

60 Dann, A.B. (1926) Wet litter in the poultry house. *Poultry Science* **3:** 15-19
61 Davies, C.N. (1951) Ventilation and its application to poultry housing. *World's Poultry Science Journal* **7:** 195-200
62 Trotter, W. (1851) Essay on the rearing and management of poultry. *Journal of the Royal Agricultural Society of England* **12:** 161-202
63 Ministry of Agriculture, Fisheries and Food (1955) *Poultry Housing*. Bulletin 56, HMSO, London
64 Holroyd, P.H. (2001) *Future Trends of Poultry Science and Practice*. 10th Temperton Fellowship Report. Harper Adams University College
65 Harrison, R. (1964) *Animal Machines*. Stuart, London
66 Brambell, F.W.R. (1965) *Report of the Technical Committee to Enquire into the Welfare of Animals kept under Intensive Livestock Husbandry Systems*. Command Paper 2836. HMSO
67 Ministry of Agriculture, Fisheries and Food (1994) *Agriculture in the United Kingdom 1993*. HMSO, London
68 Whittle, T.E. (1995) Application of science and market forces in agriculture. *Auchincruive College Association Journal* **73:** 9-12
69 Thompson, A. (1943) *Feeding for Eggs*. Faber and Faber, London
70 Richardson, D.I.S. (1976) *The United Kingdom Broiler Industry 1960-75*. Bulletin 156/EC66. Manchester University
71 Poultry World (1958) *The Poultry Handbook*
72 Hams, F.(1978) *Old Poultry Breeds*. Shire Publications Ltd., Aylesbury
73 Waldorf, E.C. (1915) Ten eggs per week per hen and how it was done. *Reliable Poultry Journal.*, Quincy, Illinois
74 Wiseman, J. (1986) *A History of the British Pig*. Duckworth, London
75 Wiseman, J. (2000) *The Pig. A British History*. Duckworth, London
76 Malcolmson, R. and Mastoris, S. (1998) *The English Pig. A History*. Hambledon Press, London
77 Hickman, T. (1997) *The History of the Melton Mowbray Pork Pie*. Sutton Publishing, Stroud
78 Long, J. (1886) *The Book of the Pig*. Upcott Gill, London
79 Cole, D.J.A. and Tuck, K. (2002) European approaches to reducing ammonia emissions. *Proceedings National Pig Waste Management Conference*, Birmingham, Alabama. In press.
80 Weir, H. (1902) *Our Poultry*. Hutchinson, London.

4

AGRICULTURE, FOOD AND THE COUNTRYSIDE: PAST, PRESENT AND FUTURE

Consider a commuter, picking up the groceries while travelling home from work. He may well suppose that there really was a relaxed and peaceful time when food production was rustic and picturesque. He may imagine that cheap, everyday commodity foods can still be produced in some 'traditional' way. Why should he suppose any different? These things are not his problem.

Does history tell us anything about the development of expectations? What historical events have forged current attitudes? Do they offer marketing opportunities for the food and agricultural industries and choices for consumers? Can these industries meet consumers' expectations? Can the countryside adapt to new expectations of it?

I. THE PAST

SOME EFFECTS OF THE HISTORY OF FOOD, FARMING AND THE COUNTRYSIDE ON EVERYDAY LIFE

When and why did occasional confrontational town and country attitudes develop? Some have dated them from the time of the industrial revolution and the associated pressure on the countryside to produce more cheap food in the first half of the 19th century.[1] This led to farmers, by then geographically increasingly separated from the enlarging towns and cities, viewing themselves as separate for the first time. The food issue was born.

The changing importance of agriculture and food: the first period of importance

The economy of Britain has not been exclusively agricultural for a very long time. Urban and commercial life, and a market economy, were well developed by the 13th century, even in rural areas. There were large urban centres by the 7th and 8th centuries and by the time of the Domesday records about a tenth of the population was living in towns.[2]

Nevertheless agriculture and rural goings on played a much larger part in the life of the nation in the middle ages, as indicated by the numbers of people employed. Until well after World War II the urban population was aware of the vicissitudes of the harvests and of the rural economy through the resulting fluctuations in the availability and price of food. Before the 20th century few towns were so large that the countryside was not nearby. From time to time some agricultural products such as wool or wheat were major export commodities. Therefore although food and agriculture did not dominate life for everyone they would have been regarded as rather conspicuous.

However the importance of the countryside in the structure of past society is underestimated by its simple economic effects. The wealth of the aristocrats, who were Society with a capital S, was based on land, and much of their income came from rents from tenant farmers. In some parts of the country the system of great estates provided the basis of a stable social order for a very long time: from the time of the Conquest until the industrialists of the 18th and 19th centuries began to make their fortunes. Many of the professional and middle classes were subject to the patronage of the land owners, and the newly rich industrialists were sometimes called 'trade' until the beginning of the 20th century. The awe in which some of the middle class held the aristocratic land owners was made fun of by the account of the sycophantic Mr.Collins, in Jane Austen's *Pride and prejudice*.[3]

Harvest festivals are still common in most parishes, despite the remoteness of agriculture from the everyday life of most of the population.

A social ritual, which has survived from the times when agricultural success mattered to nearly everyone's survival, is the harvest festival. It is still celebrated by urban and suburban church congregations, even though most of them now contain hardly anybody intimately affected by the state of the harvest. Another example is the saying of grace before formal public meals, which presumably dates from times when the simple fact of getting enough to eat was remarkable.

A harvest jug made for Roger Kendrick of Staffordshire, to celebrate the harvest of 1786.

The period of declining importance

The repeal of the Corn Laws reflected the fact that many in the burgeoning 19th century manufacturing industries, selling into world markets, saw free trade as a good thing. Protectionist agricultural policies were therefore seen as a bad thing, pushing up the price of food, and rural affairs were probably regarded as less and less relevant to the new great industries upon which the wealth of the country was coming to depend.

The second period of importance

World Wars I and II changed all that. Food supply became a very worrying matter and a priority. Its priority status continued during the post war austerity era. It must be difficult for those born in the last four decades or so to believe that food was rationed until the 1950s. Government support for agriculture, including financial support, was aimed at increasing production until well into the 1960s. There was much interest in improving the productivity of hill and marginal land, though the changes made often changed the appearance of the countryside.[70]

In today's world of consumer pull having replaced producer push it is difficult to believe how much producers and distributors held such sway over consumers in the past. Some aspects of food history now seem incredible. As recently as up to about 1850 the gross adulteration of food was so common that producers, retailers and the public all expected to put up with it.[4] When the early co-operative movement introduced pure foods some were unfamiliar enough to be rejected by consumers, because they were the wrong colour. However the Adulteration of Foods Act was passed in 1860 and the Sale of Food and Drugs Act in 1875. The 1990 Food Safety Act is very powerful and wide ranging.

Changing fashions

Fashionable attitudes changed once there was sufficient material comfort, and insulation from the effects of nature, at least for some of the population. Some see the turning point in fashion as the period of the romantic poets and artists of the 18th and 19th centuries, though I am not sure about that. Wordsworth's view of nature was picturesque but also realistic. The urge to romanticise the countryside is not new and has even been applied to the Greek countryside of classical times. The real life Arcadia region of ancient Greece was probably not particularly Arcadian.

These changing attitudes to nature and the countryside are only part of an evolving process. To prehistoric man nature must have seemed terrifying. Death by wild beasts was a real danger and so was starvation. Natural occurrences like thunder and lightning were so frightening as to seem supernatural. Through much of the historic period nature was a thing to be tamed and exploited, and the effect of the weather on the next harvest was a life and death matter.

It is not only attitudes, but also tastes which change as affluence develops. Amongst the English poor of the middle ages, and up to the mid 20th century, meat was aspirational. Meat consumption was an index of prosperity,

TABLE 4.1 CHANGING ATTITUDES TO FOOD, NUTRITION AND AGRICULTURE IN UK

Period	*Duration*	*Characteristic attitudes in UK to food, nutrition and agriculture*
Greek	to c.1650	Anecdotal and deductive nutrition and dietetics. Agricultural production underpinning society. Meat thought higher status than fruit and vegetables.
Exploratory	1492 to c.1750	Spices, novelties, exotic foods and flavours valued, sometimes for sound reasons.
Scientific	c.1650 to c.1985	Agricultural production encouraged, particularly during and after wars. Often state support for agriculture and market intervention. Inductive approach to nutrition and agricultural technology. Nutrient requirements resolved. Recommendations based on modes of action. Biochemical pathways understood. Producer push drives markets and supplies. Dairy products often fashionable and recommended ***(see consumption figures in Tables 4.2 and 4.3).*** Countryside and farming interests usually regarded as compatible.
Modern	c.1985 to date	Some signs of return to deduction. Some rejection of scientific approaches. Appeals to natural bases of nutrition. Countryside regarded as an amenity. Farming not always regarded as compatible with that amenity. Consumer pull drives markets. Consumer choice paramount. Profusion of dietary advice, both scientific and anecdotal. Fruit and vegetables fashionable ***(illustrated by figures inTables 4.2 and 4.3 below).***

and the poor complained about not getting enough of it.[4] Most fruit and vegetables, now so popular with diet writers, were seen as only good enough for peasants in the middle ages. Later, after periods of prosperity, tastes turned away from animal products. Consumers became more interested in animal welfare, food safety, environmental impact, cholesterol and saturated fats. Vegetarianism has increased. By contrast as economies develop worldwide, and as peoples emerge from deprivation, the consumption of livestock products rises.[5] Tables 4.2 and 4.3 show some trends in the consumption of animal products in UK.

TABLE 4.2 MEDIUM TERM TRENDS IN CONSUMPTION OF MEAT AND MILK IN UK IN RECENT YEARS, kg/PERSON PER YEAR

	1989-91	*1996*	*1997*	*1998*	*1999*	*2000*
Beef and veal	17.5	12.6	14.8	14.8	16.1	15.8
Mutton and lamb	7.4	6.5	6.2	6.6	6.7	6.9
Pork	13.1	13.8	14.0	14.2	14.4	13.8
Bacon and ham	7.7	8.6	8.1	7.9	7.8	8.0
Poultry meat	20.7	27.2	26.9	28.4	29.0	29.6
Milk (litres/person)	250.0	244.0	245.0	240.0	245.0	237.0

Based on DEFRA (2001a)[44]

TABLE 4.3 LONG TERM TRENDS IN THE CONSUMPTION OF MEAT AND MILK IN UK, g/PERSON PER WEEK

	1950	*1960*	*1970*	*1980*	*1990*	*2000*
Beef and veal	228	248	221	231	149	124
Mutton and lamb	154	188	149	128	83	55
Pork	9	57	80	117	84	68
Bacon and ham	128	175	177	149	118	112
Poultry	10	50	143	189	226	253
Sausages	114	103	106	92	68	60
Whole liquid milk	2716	2750	2631	2364	1232	664
Skimmed milk				*69	709	1138
Butter	129	161	170	115	46	39
Cheese	72	86	102	110	113	110

Data summarised from DEFRA (2001b)[45]
*1983, data for 1980 is not available

Data for egg consumption in India and in the UK in recent years illustrate the effect of economic background. In 1961 the consumption of eggs in India was only six eggs per person per year. By 1983 it had risen to 19, and by 1991 to 30.[6] Typically in developed countries egg consumption rises to a peak

Vegetables: once despised as food of the poor and now recognised as important components of a balanced diet.

of over 200 per person per year as affluence increases, and in the UK it reached 245 by 1965. Then, when affluence reaches a certain point consumption falls again, and UK egg consumption per person per year had fallen to 171 by 1998.[7] Another example is UK sheep meat consumption. It fell from 7.1 kg/head in 1982-84 to 5.7 kg/head in the 10 years to 1993.[8]

Dream or reality?

Was English rural life ever as romantic and idyllic as the pictures on some food packaging, calendars, magazines and Christmas cards? Does history match the dream? The answer is perhaps not always and probably not entirely; for the greater part of the inhabitants of the countryside life was often far from romantic and idyllic.

During much of the 19th century and before, the countryside for many was the scene of long hours of back breaking work for scant reward when times were good, and near starvation when times were bad, as Joseph Arch recorded from his own experiences, and as William Cobbett reported on his rides.[63,67] At the same time, by way of compensation, it is probably also true that for many life would have been tempered by a deep sense of involvement, satisfaction and belonging to the land, to the locality, to nature and to the seasons. Some of the best evocations of these attributes of old English rural employment are fictional, though probably authentic, and occur in the world of Thomas Hardy. Presumably the modern term for this compensation is job satisfaction. The concept of a rural form of job satisfaction is old. In ancient Egypt it seems that claims were made that for many agriculturalists of the time life was busy, but also dignified and peaceful.[9]

A scene in Thomas Hardy's *Tess of the D'Urbervilles* (first published in 1891) gives us a picture of the hard physical labour.[10] Some time in the 19th century Tess was working on a threshing machine of a type still in use within living memory. The passage describes the machine as a red tyrant, and its demands as despotic. (Threshing machines were usually painted pink or red.)

Tess's red tyrant, in a less tyrannical guise at a steam fair at Wymeswold, Leicestershire in 1998. (See also colour section.)

Beauty in the eye of the beholder

Ideas of the beauty of the countryside are volatile. Today many regard hedges as essential features of rural Britain and mourn their removal. But when enclosure was taking place the hedges were often regarded as symbols of oppression by those who were displaced from their land, and their planting was lamented as vociferously then as their removal is lamented now. Prothero quoted Francis Trigge (1604), who wrote that the passion for sheep and hedges had caused "merrie England" to turn into "sighing and sorrowful England".[11] By contrast other agricultural commentators saw enclosure as entirely beneficial, even to those who had become landless, so they presumably liked the new hedges for what they symbolised. "The Lincolnshire labourer living among new enclosures is well paid, clothed, lodged, and I also mention, well fed, sometimes with fresh meat".[12]

Agriculture still shapes the scenery of the countryside. Photo: S Keeling.

There is now also debate about the declining numbers of farms and the increase in size of those that remain, but as we have seen this is nothing new either.

The reality of earlier times was sometimes rather different from the modern romantic dream of the pure natural food of old. Prothero described a peasant diet of the days of the manorial system as based almost entirely on salted meat and fish, often of doubtful wholesomeness, while the livestock only survived the winter months in a state of semi-starvation.

Effects on world history

Food and agriculture have a long history of influencing the evolution of society, ever since the Palaeolithic hunters needed small co-operating groups and organised their camp sites and communities accordingly. It has even been speculated that the monogamy which is characteristic of our species originated from the need for the absent hunting male to be confident about who his children were. Then later, animal husbandry and cultivation, particularly cultivation of the cereals, made civilisation as we know it possible.

Not everyone agrees that husbandry delivered purely beneficial effects however. Cultivation led to more awareness of boundaries and frontiers, so that while agriculture might take the credit for making culture and civilisation possible, some consider that it should perhaps also take some of the blame for a predisposition to territorial disputes.[1]

Some commentators have argued that the demise of the hunter gatherer life style led to a loss of diversity of foods, with potential nutritional losses.[13] Several authors edited by Kiple and Ornelas, in their two volume history of world food, considered that the concentration of settled populations following the development of agriculture led to polluted water supplies and to diseases.[14] A historic association between crops like tropical sugar cane and slavery has been pointed out.[15]

At some periods of history food and farming have had profound and mould breaking effects on the history of the world. The medieval importance of the spice trade was a real driving force in world events, leading to exploration, colonisation and international disputes. Trading companies sometimes behaved like states. The Dutch East India Company had 200 ships and took over several countries.[16] Its English rival, the East India Company, received a royal charter in 1657 and effectively became a state, owning tracts of land in India. In the early decades of the 17th century England and Holland skirmished regularly over spices. The resulting deals over islands led to the British acquisition of Manhattan in 1667, in exchange for Run, a tiny nutmeg producing island in the East Indies.[17] Spices were so valuable that they were kept under lock and key in cupboards resembling safes. An interesting example, made in 1750, is at Rydal Mount, the home of Wordsworth in the Lake District.

Occasionally events in the history of agriculture had devastatingly unpleasant effects on the participants at the time. The Highland clearances, after 1785, and the Irish potato famine of 1845 to 1847, must surely rank as Britain's most notorious examples.[18] The former were due to the desire of the land owners to create sheep runs and the latter was due to the fungus disease potato blight (*Phytophthora infestans*) destroying what had become a staple food.[63] Both led to mass emigration to the New World, and therefore both had substantial effects on the genetic constitution and cultural background of the present populations of the USA and Canada.

Food, agriculture and English rural history

The invention of the mould board plough had profound effects on the structure of English rural society. Unlike the earlier ploughs, the mould board plough needed a team of several oxen to pull it; more than many individual families possessed. It also needed large tracts of land to operate efficiently. The benefits were so large that it motivated collaboration within villages, and thence the development of the enduring system of communal open field villages.[1]

Ironically the next major influence of land management on both society and the countryside was almost exactly the reverse. Enclosure of the large fields by the more successful of the landholders led to displacement of the others, the migration to the growing towns, and an increase in the notion of hiring out one's labour. In due course this made some aspects of the industrial revolution possible, but it also created social problems. Later still the mechanisation of field operations led to even more labour saving on farms and more migration towards industrial and urban work. This affected the appearance of much of rural Britain. Fine old churches, which would have been very expensive to build, reflecting a once large and flourishing rural population and economy, still grace villages whose modern population is tiny. This effect of mechanisation on rural work forces is a piece of history of our own times, having rapidly accelerated in the years since World War II.

Food, agriculture and commercial life

Food has contributed to the development of commercial life and of shopping in two significant ways. The branding of products is the vehicle by which consumers have trust in them. This started with food, once urban consumers could no longer know the producing farmers personally, and some of the world's most famous brands are foods and drinks: e.g. Heinz, Kellogg's, Birdseye, McDonald's, Coca Cola. Chocolate contributes five famous brands on its own: Cadbury, Rowntree, Hershey, Nestlé and Mars. Some of the leading food retailer brands are now strong in their own right. Can there be anyone in Britain who is not familiar with the names of the big four: Tesco, Sainsbury's, ASDA and Safeway? Their rise to such prominence has its own fascinating history.[19] With farm shops, WI markets and now farmer's markets, as we have seen above, we have, at least in a small way, come full circle.

The second major contribution of food to modern trading is the multiple chain store. The idea came from the Equitable Pioneers of Rochdale starting the co-operative movement in 1844 for philanthropic rather than for commercial reasons, thus giving rise to the still familiar co-op store.[4] Ownership was collective. Modern supermarkets, which are chain stores in the Rochdale tradition, have given consumers choice and value for money, as well as round the clock convenience, though there may be segments of the population unable to take full advantage of this because they live in the wrong places, or lack access to a car.[19]

W.I. markets started in 1919 and are still popular events. Photo by permission of the National Federation of Women's Institutes.

Farmers' markets are growing in popularity. This is Melton Mowbray farmers' market.

Agriculture and its contribution to other technologies

The need to measure parcels of land gave birth to mensuration, and the need to carry out field operations at the right times of year led to the beginnings of astronomy and to the development of the calendar. It is interesting that such a fundamental subject as mathematics started for such practical reasons.

The Egyptian civilisation provides examples of the early history of mathematics. For administrative reasons, including food rationing and the provision of rewards and salaries in an economy without money, the Egyptian scribes devised an arithmetic system and some weights and measures, mostly based on grain. Yields of grain were measured in scoops of known capacity in *hekats* (1 *hekat* was about 4.78 litres). They even developed an arithmetic system for coping with circular containers of grain and other products, and therefore they had an equivalent of the later Greek π. The unit of linear measure was the *cubit* (523 mm) subdivided into units based on fingers and palms. A *temple day* was one 360th of a year.[20]

Obsessions with food in the past

Obsession with food is not new, and nor are changing fashions in ideas on healthy eating. The ancient Egyptians had some strongly held views on the subject.[1] Up until about 1650 advisers on food matters made recommendations based on Greek and Roman ideas on balancing body fluids.[21] Aristotle considered matter to be hot, cold, wet or dry, and Galen had classified body fluids as blood, phlegm, yellow bile and black bile.[22] These four humours were thought to need balancing, and certain foods encouraged each of the four, so that the cook and the physician were advised to adjust the menu to achieve the supposed balances. Digestion was thought to be a form of cooking. Some modern food fashions are scarcely any less bizarre.

After about 1650, under the influence of Paracelcus, Galen's four humours were replaced by three: volatile, fluid and oily. Digestion was considered to be a form of fermentation (not far from the truth in the case of ruminants, horses and pigs, as it happens). Menus changed to include foods believed to be readily fermentable, and items like mushrooms came back into acceptability. Sugar was perceived as dangerous, and of mainly decorative value relegated to separate dishes, hence the appearance of dessert at the end of a meal.[21]

II. THE PRESENT

CAN EXPECTATIONS BE MET?

The social and economic significance of food

Food manufacture and food retailing are now big business and important sources of employment. At late 1990s values the world's largest food manufacturer was Nestlé. Food retailing is also big business, with the sales of Tesco, for example, at £25.7 billion in 2002.[19, 23, 62]

Modern food retailing involves presentation and choice. Photo: Tesco.

The social significance of food has been substantial in Western culture and in most other cultures; and it still is.[24] Shared eating has long had importance for bonding and social cohesion, so much so that there was concern after the popularisation of fish and chip shops that they would be socially corrosive. There is now a similar concern that modern fast food and snacking habits might be socially destabilising and contrary to the interests of family life. Food has deeply rooted functions in ritual; both religious and secular. Gofton pointed out, for example, that the word carnival is derived from *carne*-val, because of associations with meat.[24] Literature reflects the social importance of food.[25] In Jane Austen's novels many of the famous scenes, such as the picnic on Box Hill, are occasions where the characters come together to eat. We still eat together for the strengthening of family, social and business bonds.

Concerns, scares and some statistics

Food and agriculture seem to be particularly prone to finding themselves at the centre of modern debates on the suspicion of science; and even of much wider issues like the globalisation of business, capital, commerce and trade.[26]

Modern retailing offers consumers a wide choice. These pictures show, for example, healthy options and prepared foods. Photos: DJA Cole.

For many, though by no means for all, in modern Britain food is no longer a dominant part of the household budget. It is now a source of interest and pleasure as well as sustenance. It has therefore also become a popular talking, writing and TV discussion point. For some it has, perhaps, become something to take some of the blame for life failing to meet zero risk expectations.[27]

Food safety and environmental concerns frequently feature in news stories. During 1998 the media covered 14 food scares according to my records, though I may have missed some. In the last few years food safety and healthy eating concerns have accelerated the growth of consumer scepticism. There is now a long list of products which have, either routinely or occasionally, been the subject of food scares. Hygiene and food poisoning have been the highest profile scares but many foods have been in the news when reported to contain carcinogens. One of the latest debating points is genetically modified foods. In response to public concern about some of these matters the Food Safety Act of 1990 has wide powers, and the Food Standards Agency was established in 2000.

There is little doubt that producers and distributors must take these concerns seriously, since it is consumers who drive today's market.

There have been occasional voices of caution about the true scale of the problem, in particular in a book edited by Bate in 1997.[28] The claim that modern technology leads to less safe food, or less environmentally friendly food, has been challenged by several authors edited by Morris and Bate.[29] They posed questions such as how many fewer would have died of famine had regulatory intervention been less severe? The claim that things are generally getting worse in the world has been famously questioned in Lomborg's recent book.[30] P.J. O'Rourke posed similar questions more amusingly if less politely.[31]

RETAIL AND CONSUMER POWER, AND THEIR EFFECTS ON THE FOOD MARKET AND THE COUNTRYSIDE

The British countryside is much talked about these days, but often as a backdrop for living and leisure, rather than as a food source. Some have chosen to live in rural areas and to commute to work, and for these the local farms are important, but more for their appearance than for their productivity. As an interesting aside it seems that the classical ancient Greeks commuted in reverse, choosing to live in the *polis* (city), where presumably the political and cultural action was, and travel out to their farms.[32]

For the modern Briton striving to make a living from industrial or post-industrial activities, and heir to many generations of urban and suburban lifestyles, the agricultural use of land is understandably not always a high priority. Consumers are customers of agriculture and they have their own preoccupations, problems and priorities. It is useful for producers to recognise that this is not new, and some of history's pioneering farmers understood it well. When Robert Bakewell carried out his improvements to the cattle and sheep of his day he did so in response to a market opportunity which he had been shrewd enough to spot. His public, and he used that word, at the time wanted fat meat.[33] Effectively he had done his market research, though it is doubtful if the term would have been part of his vocabulary.

A legacy of the priority period is that it has not always been easy to adjust to new requirements. In today's Britain the food supply chain is not dominated by the needs of producers but by the dictates of consumers. The public have come to expect that competitively priced, interesting, high quality, attractively presented, safe, convenient and often ready prepared food, including items from all over the world, is available and offered for sale seven days a week, round the clock. Consumers are now sometimes more concerned about the countryside than about the production taking place in it. Food production businesses and sectors wishing to prosper have had to adapt to these changed attitudes. Some of these changes are illustrated by the proportion of the final value of food now added after it leaves the farm gate. They are also indicated by the small number of people employed in primary agriculture compared with the several million working in the whole food chain.

The appearance of the countryside is an issue nowadays.

It is not just producers who have sometimes experienced difficulties of adjustment, but professionals also. In 1995 Blaxter and Robertson published a history of the improvements which had taken place in food production and supply, and of the scientific and technical achievements making them possible, in a book entitled *From dearth to plenty: the modern revolution in food production.* In his introduction Sir Kenneth Blaxter, writing at the end of an illustrious career which included substantial achievements of his own, recorded that the book had been conceived in both pride and anger.[34]

Nevertheless changes have happened, and will continue to happen, so market research and adaption will continue to be appropriate.

A socio-economic perspective affecting the food market

There is much talk of affluent consumers driving the changes in food requirements in today's UK, and this is true, but it is not the whole truth. Statistics for 1999/2000 from the Office for National Statistics suggest that the disposable weekly household income of the lowest income quintile (*i.e.* 1/5 of the population) was only £94. Of this over 20% was spent on household food. The corresponding income figure for the highest quintile that year was £882. It is apparent which groups might be most likely to constitute the main markets for commodity foods and which groups might be the main purchasers of premium food products. The higher income groups spent a lower percentage of their income on food, but more in £ per household.[35,36]

Trends in household income over time have been analysed by the Office for National Statistics. There was little change in the distribution between household income percentiles during the 1970s or the 1990s, but changes occurred during the 1980s, when the incomes of the higher income groups increased faster than those of the lower income groups. During that period the median "real household disposable income", adjusted for inflation, rose by 27%, that of the 90th (i.e. top) percentile rose by 38%, and that of the 10th percentile by 7%.[37]

Markets and rules

At this point it might be worth a brief digression to explore the historical perspective of modern food and agricultural rules. Often the procedure is that the EC makes directives and then member states are obliged to formulate corresponding regulations; though member states may choose to make additional legislation, codes of practice and product certification rules. Food and agriculture related topics covered by directives, regulations and other rules these days include food safety, animal welfare, environmental issues and organic certification.

However the retailing and catering supply trades are now so competitive and responsive that effectively it is they who tend to make some of the rules about food. If consumers do not want something they do not buy it, and that message is very quickly received and acted upon by retailers and caterers. Equally if consumers really want something special, and they are prepared to pay a higher price for it, then it becomes an opportunity for the trade and it will probably quickly appear on the market.

We have seen that this has been happening throughout the history of food ever since the first cheese appeared. It is called niche marketing and it can be a good thing for both customers and suppliers. Recent (1980s and

1990s) examples are well known: free range eggs; Freedom Food (run by the RSPCA); organic produce bearing the trade marks of the certification organisations such as the Soil Association; meat from rare breed livestock; game; wild boar; wild mushrooms; venison; and ostrich; to name but a few.

Many convenience and factory made foods have been considered by some to be a legitimate part of the dietary heritage of an industrial country like Britain. A long tradition of an industrial food industry has led to useful innovative skills for today's needs for added value.[38]

Speciality, local and traditional foods

There is now increasing consumer interest in the heritage of food and in traditional, regional and speciality foods.[39]

Food now provides variety and interest as well as nourishment. Geographical isolation in historic times has given us a rich variety of local and regional foods in Britain. Above are some of the hundreds of cheeses to be found. Photo from Middle Farm, Firle, Sussex, by John Pile ©2002 Middle Farm Limited.

A fairly new development at the time of writing is farmers' markets, though they are not dissimilar from the Women's Institute markets, which date from 1919, and which were mentioned by Stapledon in 1944.[40,41] These enable producers to sell direct from stalls in towns, and they enable consumers to recognise the individuals who have produced the food, thus returning to the traceability and sales methods of earlier times. All these developments offer producers new opportunities and they offer consumers traceability, interest and choice.

Melton Mowbray pork pies: an example of a regional speciality. Photo provided by F. Bailey and Son, family butchers, Upper Broughton, Melton Mowbray, who have been making pies at that address since 1905[71].

Surveys and reality

The real test of what consumers want is what they buy, which is not always the same as what they say they want. Reay Tannahill quoted several examples of surveys where interview results did not agree with sales statistics.[1] In one example 63% said that they had increased their consumption of fruit and vegetables at a time when measured consumption fell. In a study in Tucson, Arizona, 85% said that they did not drink beer, but a sift through their dustbins revealed that 75% had beer cans in them.

History as affecting present issues and attitudes to food, agriculture and the countryside

The story told in the chapters above shows how the history of food and agriculture could be said to have driven the history of the world. But how has that history affected present day British food and countryside issues?

Attitudes to food, nutrition, agriculture and the countryside seem to have had a history of their own. Four eras can perhaps be identified, and they are speculated in Table 4.1 on page 113. They overlap, of course, as there is seldom a perfect consensus of opinions, least of all in topics with emotional overtones. Current attitudes to food have also been classified according to, among other things, degrees of involvement and rationality of choices.[42,43]

During most of the period of recorded history in Britain, for most of the people, all that was required of food was that it should be available, affordable and nourishing; though the better off have always wanted something more. This affected attitudes to agriculture and the countryside, even for those not involved in food production. But until quite recently much larger proportions of the population were involved.

As a result of some of the social changes described in earlier chapters, and of the changing attitudes postulated in Table 4.1, different attributes are required of modern food and of its production and supply. They could be summarised as in Table 4.4, not necessarily in order of importance, but classified as biological needs or wants. As individuals, families or countries gain in affluence, wants assume more and more importance. (Needs are things without which a healthy life is not possible. Wants arise from a wish to make life more interesting.)

TABLE 4.4 SOME ATTRIBUTES OF FOOD AS NEEDS AND WANTS

Needs	Nutrition, satiety, food safety, affordability, availability, reliability
Wants	Palatability, presentation, aroma, suitability for cooking, convenience, naturalness, entertainment, emotional value

Food marketing specialists recognise emotional values of food, some of which have historical and cultural origins. Table 4.5 offers a list.

TABLE 4.5 SOME EMOTIONAL VALUES OF FOODS

Emotional attribute	*Reference*	*Countryside relevant or not*
Heritage and story of the food	Hughes and Ray (1999)[46]	Yes
Novelty, variety	Van Trijp and Steenkamp (1998), quoting Hughes (1994)[47,48]	Sometimes
Packaging, presentation		No
Provenance, authenticity	Moore (1993)[49]	Yes
Farming system origins		Yes
Purchase experience		No
Retail experience and convenience	Seth and Randall (1999)[19]	Perhaps, if retail outlet availability affected. Yes in farmers' markets and farm shops
Social and other associations	Gofton (1996)[24]	Sometimes
Naturalness		Sometimes

Historically, and particularly in times of war, the expectations placed on agriculture and food have emphasised supply security; therefore markets tended to be producer led. Now, as in the late 20th century, markets are consumer led. "*Producer push* has now given way to *consumer pull*".[50]

The balance of pressures and influences on the food industry and the countryside

Modern food trading is complex. It is affected by both global and national pressures. Figure 4.1 summarises some of them.

Recent analysts have stressed the effects of forces such as globalisation on the need for more structuring in agriculture, as well as for devices like collaboration and partnerships. It has been suggested that farm level agricultural support has a negative effect on competitiveness. Protectionist policies are coming to an end, and are being replaced with limited social and environmental support.[51]

Changing expectations of the countryside

The changes described above probably help to account for changing expectations of the countryside, from its role as a food production base to a relatively new role as an amenity. The changes may, perhaps, be illustrated by a non-food example. When some of the uplands and marginal lands were planted with rows of conifers, for efficient production of scarce timber, it seemed a sensible thing to do because the expectations placed on the countryside were biased towards productivity and supply. Now that the trees have grown, as they were intended to, such plantations are frequently criticised because the expectations have changed: there is now a stronger wish for the countryside to supply scenery than to supply timber. Value judgements are not the point of such an example; merely the recognition of changing times.

In fairness to the writers of the times when the conifers were being planted, it must be acknowledged that they did not all fail to forecast what were then very long term demands on the countryside. Sir George Stapledon first published *The Land Now and Tomorrow* in 1935, and in several further editions until 1949. Writing as long ago as that he correctly forecast that by 1999 people would have time to enjoy the countryside and to spend money in it. He reminded his readers that the terms of reference of the Forestry Commission were to produce timber, though in the same chapter he also advocated what he called purely aesthetic planting, and regretted that when

Changing uses and expectations of the countryside. This example of a new aspect of the countryside is a wild flower farm at Langar, Nottinghamshire. Photographed with the permission of Naturescape.

the Forestry Commission was established in 1919 it was only timber which was considered. Stapledon foresaw some of the changes in emphasis which eventually led to changing the Ministry of Agriculture, Fisheries and Food (MAFF) to the Department for Environment, Food and Rural Affairs (DEFRA) and to some of the functions of the Countryside Agency. He outlined what national parks might become and envisaged a Ministry of Lands, recognising the contributions of the land, other than food production, to the national life.[41] National Parks as we now know them date from 1951.

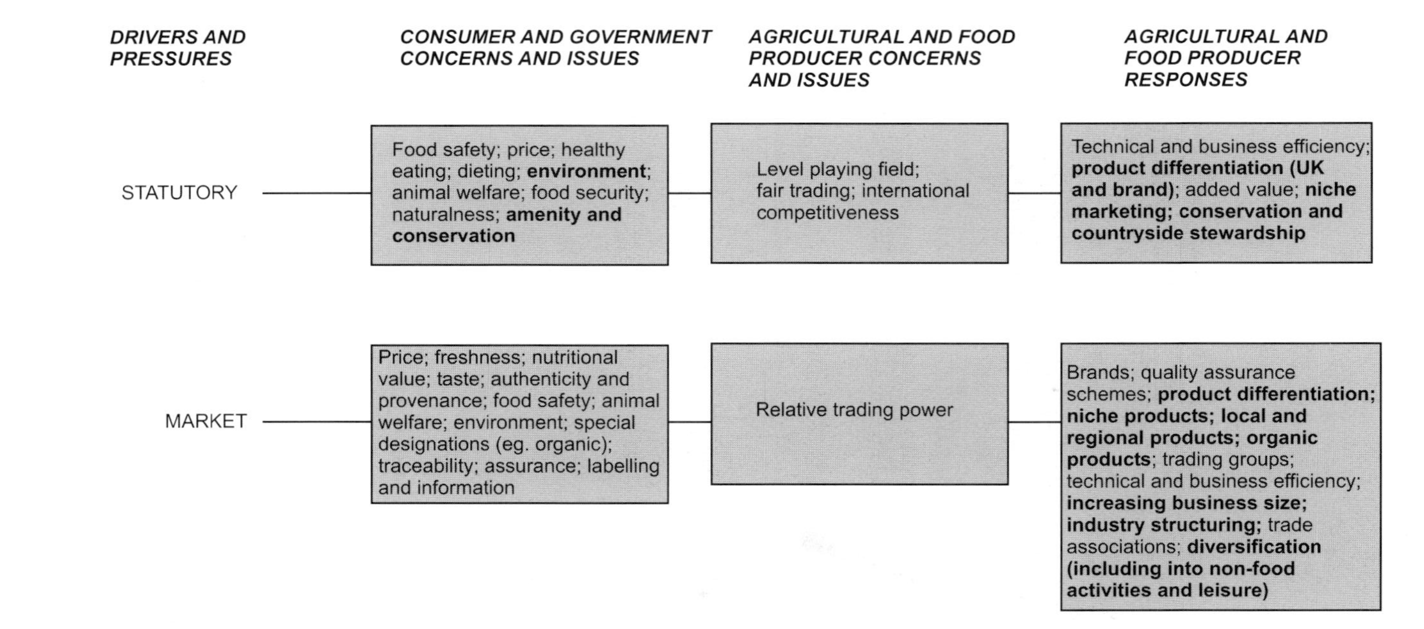

Figure 4.1 Some pressures and issues in the food trade, not necessarily in order of importance (items in **bold** are those which may also affect, or be affected by, countryside issues).

Some of the countryside is used for activities other than agriculture.

His national parks were to contain working farms and therefore rural employment. This view of the countryside as a work place is still current. Most of the organisations described below as having an interest in the countryside want to see employment in rural areas as part of a living countryside. There will probably be a difference between the rural employment of the past and of the future, however. Whereas throughout history most of the rural employment was in primary agricultural production of basic commodity products, future employment may be more varied. Agriculture will continue to play its central part in the life and functioning of the countryside, but there is likely to be a far broader pattern of occupations, and of products, than in the past.

Another illustration of changing fashions in expectations of the countryside occurs in much older writings than those of Stapledon. Moors and heaths are now valued as wildlife habitats and places for recreation, but they were once seen as evidence of their owners failing in their duty to produce. *Baxter's Library of Agriculture* of 1846 quoted a Mr. Pusey who, in 1842, had described a successful drainage project as "a beacon to encourage landlords in converting their dreary moors into similar scenes of thriving industry".[12]

The countryside as a place in which to live

The concept of the countryside as primarily a place in which to live or to visit, rather than as a food production base, is now part of the commuting way of life. As a result, the rate at which English villages have become suburban has

Cowslips in grass.

The countryside as scenery.

accelerated in the last few decades; though perhaps less so in the remote areas. Many local historians have documented the evolution of their own villages. Two recent examples, from different parts of the country, were in 1994 in Nottinghamshire, and in 1997 in Herefordshire.[52,53] Such publications remind us that up until the 19th and early 20th centuries the majority of people living in villages made their living in the same villages, as rural populations had done for centuries. That all changed in the latter half of the 20th century, particularly

By the turn of the 21st century diversification was an important approach to increasing farm incomes. At this rally of vintage tractors in south Nottinghamshire, in August 2000, a float featured farm holidays.

after the car made commuting easier. Interestingly a modern social change, re-introducing the idea of living and working in the same place, including rural places, is the electronic office and home working. This book is being written at a rural location and e-mailed to the publisher in another rural location.

Examples of changes in English villages

The working life of the countryside is still dominated by agriculture, but there have been changes in the other occupations taking place in rural areas. In former times most of the other village occupations were in support of agriculture. Most villages had one or more farriers, blacksmiths, wheelwrights, carters, bakers and pubs. Most still have the pubs but not the other trades. Many of the pubs are now also restaurants, often with children's play areas. Therefore the uses of village buildings have changed and new buildings have appeared. The offices of the publisher of this book (Nottingham University Press) provide a typical example. They are in a former farm building in Thrumpton, south Nottinghamshire. Thrumpton is an example of a village which, though still very small and rural, now has several non-agricultural commercial enterprises in addition to its farms.

By contrast Keyworth, also in south Nottinghamshire, might serve as an example of villages in the English Midlands which are no longer small. Keyworth is fortunate in having been the subject of some scholarly publications by its local historians.[52] It still has farms, but their numbers fell from 35 in 1894 to 11 in 1984.[54] Over the same period the population increased, slowly at first, and then rapidly during the 20 years from 1951when the village was becoming a commuter base, mostly for people working in nearby Nottingham.[55] The population in 1800 was less than 500 and by 2001 it was 7,420, though there was a boundary change in 1984.[55,56] As well as the pubs there are now shops, including supermarkets, and all the usual facilities of a large village or small town, such as a library, a health centre, some small industrial units, and several schools. There is one major employer: the British Geological Survey. Interestingly in the present context there is also a small nature reserve. There must be numerous dormitory towns and villages in Britain with similar stories to tell.

Expectations of the countryside

The perception of a separation of food production from the countryside has changed public expectations of it. Whereas a century ago, or even 50 years ago, a typical citizen, particularly in the rural areas, would have expected the countryside to be attractive, but primarily productive, current expectations might sometimes rate productivity much lower in priority. Some of the modern expectations of the countryside might be listed as including: amenity, accessibilty, visual appeal, space, wild life habitats, resemblance to a perceived ideal, smell free, not too crowded; but yet not too deprived of modern conveniences. Can all this be realised?

Keyworth old and new. Photos: Nigel Morley and Keyworth and District Local History Society.

Some statutory, charitable and non-government organisations

The variety and tenor of statutory and charitable organisations with interests in the countryside is an indication of current public concern for it.

Changing public concerns have led to changing official bodies. The Ministry of Agriculture, Fisheries and Food (MAFF) became the Department for Environment, Food and Rural Affairs (DEFRA) in 2001. In earlier chapters we have seen the activities of MAFF and its predecessor, the Board of Agriculture, in supporting production throughout a long period when that was the priority. The changes of name suggest the recent changes in emphasis and in public concerns.

Modern concerns about the countryside have stimulated the development of other official bodies with emphasis on it. The Countryside Agency, partly known as the Countryside Commission until 1999, provides support in several areas of rural conservation and development. It is responsible for designating National Parks and Areas of Outstanding Natural Beauty, and it produces the Country Code for guidance on the treatment of the countryside as a public amenity. It has operated schemes promoting the link between food and countryside, such as *Countryside Products*, and later *Eat the View*. Its Countryside Stewardship project has been widely adopted.

The National Farmers' Union (NFU), formed in 1904 in Lincolnshire and going national in 1908, represents the interests of farmers, as its name suggests. There is also a Tenant Farmers' Association. The Country Landowners Association (CLA) was formed in 1907 as the Central Land Association, becoming the CLA in 1949. Its new title of the Country Land and Business Association reflects current changes in rural business activity. Since before World War II there have been local and county Young Farmers' Clubs, and these now have national and county federations and international connections. The heritage of the union for agricultural workers goes back to Joseph Arch in 1872.

English Nature was set up in 1990 under the Environmental Protection Act and is DEFRA funded. It supervises Sites of Special Scientific Interest (SSSIs) and National Nature Reserves (NNRs). It implements EC directives such as the Conservation (Natural Habitats) Regulations 1994.

There are voluntary and charitable organisations concerned with the care of the countryside and its heritage, among other things. The National Trust, founded in 1895, is a land owning example. The Council for the Protection of Rural England (CPRE) is a promotional and campaigning group, which also prepares specialist studies and publications. English Heritage looks after historic buildings and sites, but also has an interest in landscape.

Another organisation with a distinguished history of contributions to the social infrastructure of the countryside, and of campaigning on countryside and agricultural issues, is the Women's Institute (WI). Its foundation was in Canada in 1897, reaching Anglesey and then England during 1915. Its produce markets have been mentioned several times above.[40]

World War I

World War II

World War II

The WI and the war effort. Jam making and fruit picking scenes. Photos by permission of the National Federation of Women's Institutes.

The Church of England can claim a long association with the administration of the countryside, ever since the 6th or 7th centuries. In the late 6th century Pope Gregory corresponded with Archbishop Augustine about endowments, and it has been claimed that Augustine organised a system of parishes. Glebe land was allocated for the benefit of the clergy. To this day church towers and steeples dominate the landscape and mark the location of parishes, as they have since the times of the open field system. In 1990 the archbishops produced a report on the countryside, making 47 recommendations, of which 24 were secular, and concerned with social, environmental and infra-structural aspects of the countryside. Issues such as the role of rural shops, post offices and transport were discussed.[65,66,67]

For centuries the church was associated with the administration of the countryside.

Other organisations interested in the countryside and landscape include the Woodland Trust, which aims to influence the future of native woodland and owns some of it. There are others whose aims are not primarily about the countryside, but whose activities affect it, such the Royal Society for the Protection of Birds (RSPB), the county wildlife trusts, the game conservancy interests and some of the outdoor sporting and leisure interests. The Royal Society for the Prevention of Cruelty to Animals (RSPCA) is a group not primarily concerned with the countryside, but whose interest in farm animals affects countryside issues.

The Countryside Alliance represents country pursuits, including field sports, and runs food and farming projects, including produce sales and box schemes.

This list is probably far from comprehensive.

III. THE FUTURE

Most of this book describes events of the past. Some of the above remarks in this chapter describe the present. Those are the easy bits.

To attempt to extrapolate from the past and the present into the future is much more difficult. Prediction is a hazardous business, partly because in agriculture, food and countryside change is so rapid. The importance of change in these topics is so great that Heraclitus (513 BC) was recently quoted at an industry conference for his observation that nothing is permanent but change.[57] Given such difficulties, this book can only offer observation and speculation, not certainty; still less does it purport to offer advice or recommendation.

Changes are, if anything, quicker than ever now; because more agricultural products are sold into free markets and on increasingly global markets: demographic and socio-economic changes are rapid. The rapidity of change in free market agricultural products has been illustrated by the recent history of the eggs and poultry markets.

In view of the difficulties, particularly with quantitative prediction, some published attempts at forecasting have wisely taken into account the possibility of changing influences on the factor under consideration. The prediction of change is a discipline in its own right, with its own methodology. Assumptions have to be made on factors such as the shape of the trend line to be used. Linearity is simple and tempting, but has sometimes led to some strange predictions. The further ahead in time for which the forecast is attempted the less secure are the projections. Thus, by way of explaining these difficulties, nine different projections have been made for the world population by 2050. The highest estimate (15.0 billion) was nearly twice the lowest (8.3 billion), the 1997 population having been taken at about 5.8 billion.[58]

An attempt to examine the possible trends for agriculture, food and the countryside might, perhaps, begin with a résumé of some current issues likely to remain important. In the UK and the EU that probably includes most of those in Table 4.4. In addition, the consensus of food chain speakers at a strategic conference in 1997 was that three things ranked high amongst their concerns: they were consumer satisfaction, information and food safety.[59] At a food chain event three years later the speakers urged attention to the need for trust throughout the food chain, the value of partnerships, the need for communication and the importance of information.[60]

As protectionist policies come to an end there will almost certainly be increased economic pressure on commodity products, particularly those which compete with produce from countries with inherently lower costs. Niche and speciality products have been appreciated by consumers, and therefore profitable for some producers and companies, but by definition these products cannot account for a large proportion of the total food market.

Since the food trade is global, and UK prices are affected by world events, it would be inappropriate to discuss the future in the UK without taking the rest of the world into account.

Debates about the ability of the earth to support the growing world population have raged ever since Malthus (1766-1834). Current predictions tend to be a balance between pessimism on the part of environmentalists and ecologists, and the optimism of economists and agricultural scientists.[58] It is interesting that in his appraisal Cole included water and food in the same chapter. He identified 81 possible outcomes for the food supply by 2050, depending on combinations of population growth, land use, yields and trading conditions.

Unexpected events have a habit of introducing even more uncertainties, and the perturbations to trend lines are often large. At several points during history we have seen the effects of wars on the economics of agriculture and food, and on policy; but wars are not the only causes of unexpected perturbations to gradual change. An event which rocked economic predictions which had hitherto seemed robust was the energy crisis of 1974, since modern food production is energy dependent.

At this point let us briefly consider the energy cost of UK agriculture. The consideration is real and important, but there are some interesting perspectives to be borne in mind. Firstly, on the one hand, its importance is relative. In 1999 agriculture accounted for only 0.6% of total inland energy consumption.[36] This may be compared with, say, air transport at 4.7%, road transport at 17.9%, or domestic uses at 20.2%. However, on the other hand, this is only the direct energy usage. Full impact analyses, including indirect use, may yield different answers. The energy use represented by a tractor, for example, is not just the fuel used while it is in field work: it is also the energy cost of its manufacture; everything from smelting the metal onwards. Should we ascribe the energy cost of manufacturing such a product to the agricultural sector or to the manufacturing sector? Corresponding appraisals of the indirect energy costs of capital equipment apply in many other industries of course.

The proportion of the population working in agriculture is now small in the UK, and so is its contribution to the gross domestic product (GDP). Therefore it seems probable that in the absence of some reason for food security assuming a higher priority than it has had since the ending of rationing in 1954, the public may continue to regard the countryside more as an amenity than as part of the national industrial base. This raises questions about countryside stewardship and its financing, about its management, and about objectives for it.

	Agriculture, horticulture and food production	What is the appropriate balance between the future roles of farming? Growing role for added value products and rural food manufacture?
	Tourism, amenity, gardens and parkland	Who pays for maintenance? Right to roam?
  	Leisure and sports, including: walking, rambling, hunting, shooting, fishing, other country and field sports, camping and caravans, golf courses, theme parks and heritage, equine activities, bird watching, nature, picnics, motoring, cycling, sketching and painting, boating, climbing, etc.	Who pays for maintenance? Right to roam? Where will roads and parking be located? What services will be needed (catering, toilets and seating)? How will these affect the quality of the scenery? Can confrontations be avoided and can interest and lobbying groups be reconciled?
	Natural habitats and reserves	Who pays for maintenance? Relationship with farming (hedgerows etc.)?
	Forestry	Commercial or amenity priorities?
	Residential	Planning and location questions? Rural transport questions. Balance between public transport and private cars? Survival of village shops and post offices? Effects on the elderly, the very young and the immobile? Effects on demographic balances (e.g. age distributions in rural population, retirement properties)? Rural services, utilities and infrastructure? Future of village halls?
	Non-agricultural employment	Planning and location questions

Some issues and debates which will probably be of concern during the next few years.

Under the heading of objectives, a brief historical digression on food security may be of interest. The medieval city fathers regarded it as a city by city issue. They wanted a secure supply, particularly of bread and meat, for their town and they imposed trading rules in an attempt to achieve it. They even fixed prices.[61] In the wars of the 20th century food security mainly meant having enough UK home food supply so as to be not too dependent on shipping and imports. In the age of globalisation is it that simple, and is it less of a country by country problem or is it more? How serious do balance of trade problems have to be to count as emergencies?

Visionaries of the past have addressed issues and objectives for the countryside, sometimes with remarkable foresight. An example quoted above was Stapledon in the 1930s and 1940s, but there have been many others. The stories told in the preceding chapters are full of inspirational visionaries.

Thus despite the hazards of attempting projections there are two forecasts which seem fairly safe to make. One is that the next few years and decades may see changes as significant as those described in earlier chapters. Some of the changes will probably be structural and some could potentially affect the countryside.

The other safe forecast is that coping with these changes will depend on an adequate supply of two key requisites: information and visionary people. Perhaps an understanding of the history of what has gone before may help to prepare the visionaries and the rest of us for what is to come?

Footnote:

This book is primarily concerned with the influences and trends of the food and agricultural past, which naturally raise questions about the future, and which should help us with an understanding of some principles and perspectives for the future. However countryside, land use and landscape are disciplines with an extensive literature of their own, and they have specialist professionals working on them. Therefore some further reading is listed below, after the main list of references. The further reading list is not comprehensive; it is a small sample of what is available. There are also specialist journals and periodicals for these disciplines.

REFERENCES

1 Tannahill, R. (1988) *Food in History*. Penguin, London

2 Dyer, C. (1994) *Everyday Life in Medieval England.* Hambledon Press, London

3 Austen, J. (1813) *Pride and Prejudice*. Fontana edition 1980, Glasgow

4 Burnett, J. (1989) *Plenty and Want. A Social History of Diet in England From 1815 to the Present Day*. Routledge, London

5 Thomas, P.C. (1998) Key note lecture. *32nd University of Nottingham Feed Manufacturers Conference*, Edit. Wiseman, J. and Garnsworthy, P.C., Nottingham University Press

6 Panda, B. and Mohapatra, S.C. (1993) Poultry development strategies in India. *World's Poultry Science Journal* **49:** 265-273

7 Ministry of Agriculture, Fisheries and Food (1999) *Agriculture in the United Kingdom 1998*. HMSO, London

8 Ministry of Agriculture, Fisheries and Food (1994) *Agriculture in the United Kingdom 1993*. Stationery Office

9 Aldred, C. (1961) *The Egyptians*. Thames and Hudson, London

10 Hardy, T. (1891) *Tess of the d'Urbervilles*. Penguin edition 1978

11 Prothero, R.E. (1936) *English Farming Past and Present*. Heinemann and Frank Cass & Co., London

12 Baxter, J. (1846) First published 1830. *Baxter's Library of Agriculture*. Fourth edition.Lewis, Sussex

13 Crawford, M. and Ghebremeskel, K. (1996) The equation between food production, nutrition and health. In: *Food Ethics*. Edit. Mepham, B., Routledge, London

14 Kiple, K.F. and Ornelas, K.O. (2000) *The Cambridge World History of Food*. Cambridge University Press

15 Musgrave, T. and Musgrave, W. (2000) *An Empire of Plants. People and Plants That Changed the World*. Cassell, London

16 Toussaint-Samat, M. (1994) Translated Bell, A., *A History of Food*. Blackwell, Oxford

17 Milton, G. (1999) *Nathaniel's Nutmeg*. Hodder and Stoughton, London

18 Prebble, J. (1963) *The Highland Clearances*. Penguin, London

19 Seth, A. and Randall, G. (1999) *The Grocers. The Rise and Rise of the Supermarket Chains*. Kogan Page

20 Kemp, B.J. (1989) *Ancient Egypt*. Routledge, London

21 Lauden, R. (2000) Birth of the modern diet. *Scientific American* **283:** 62-67

22 Russell, B. (1984) *A History of Western Philosophy*. Unwin, London

23 Kotler, P., Armstrong, G., Saunders, J. and Wong, V. (1999) *Principles of Marketing*. Prentice Hall, N.J.

24 Gofton, L. (1996) Bread to biotechnology: cultural aspects of food ethics. In: *Food Ethics*. Edit. Mepham, B., Routledge, London, 120-137

25 Lane, M. (1995) *Jane Austen and Food*. Hambledon Press, London

26 Micklethwaite, J. and Wooldridge, A. (2000) *A Future Perfect. The Challenge and Hidden Promise of Globalistion*. Heinemann, London

27 Dalrymple, T. (1998) *Mass Listeria - the Meaning of Health Scares*. André Deutch, London
28 Bate, R. (1997) *What Risk?* Butterworth Heinemann, Oxford
29 Morris, J. and Bate, R. (1999) *Fearing Food. Risk, Health and Environment.* Butterworth Heinemann, Oxford
30 Lomborg, B. (2001) *The Skeptical Environmentalist.* Cambridge Univeristy Press
31 O'Rourke, P.J. (1994) *All the Trouble in the World.* Atlantic Monthly Press, New York
32 Alcock, S.E. (1998) Environment. In: *Ancient Greece.* Edit. Cartledge, P., University Press
33 Stanley, P. (1995) *Robert Bakewell and the Longhorn Breed of Cattle.* Farming Press Books, Ipswich
34 Blaxter, K.L. and Robertson, N. (1995) *From Dearth to Plenty. The Modern Revolution in Food Production.* Cambridge University Press
35 Office for National Statistics (2000) *Family Spending.* Stationery Office
36 Office for National Statistics (2001) *Annual Abstract of Statistics.* The Stationery Office, London
37 Matheson, J. and Babb, P. (2002) *National Statistics. Social Trends.* No.32 Stationery Office, London
38 Ellis, H. (2001) *Eating England.* Octopus Publishing, London
39 Mason,L. and Brown, C. (1999) *Traditional Foods of Britain.* Prospect Books, Totnes
40 Goodenough, S. (1997) *Jam and Jerusalem.* Collins, London, and National Federation of Women's Institutes
41 Stapledon, R. G. (1944) *The Land Now and Tomorrow.* Faber and Faber, London
42 Meulenberg, M.T.G. and Viaene, J. (1998) Changing food marketing systems in western countries. In: *Innovation of Food Production Systems.* Wageningen Pers, 5-36
43 Grunert, K.G., Baadsgaard, A., Larsen, H.H. and Madsen, T.K. (1996) *Market Orientation in Food and Agriculture.* Kluwer Academic Publishers, Boston (quotation only seen)
44 Department for Environment, Food and Rural Affairs (2001a) *Agriculture in the United Kingdom 2000*, Stationery Office, London
45 Department for Environment, Food and Rural Affairs (2001b) *National Food Survey 2000.* Stationery Office, London
46 Hughes, D. and Ray, D. (1999) *Developments in the Global Food Industry. A Twenty First Century View.* Wye College, University of London
47 Van Trijp, J.C.M. and Steenkamp, J.E.M.B. (1998) Consumer oriented new product development: principles and practice. In: *Innovation in*

Food Production Systems. Edit. Jongen, W.M.F. and Meulenberg, M.T.G., Wageningen Pers, 37-66

48 Hughes, D. (1994) *Breaking with Tradition. Building Partnerships and Alliances in the Food Industry*. Wye College Press (quotation only seen)

49 Moore, J.M. (1993) Safety and quality of food from animals: the consumer's view. In: *Safety and Quality of Food From Animals*. Edit. Wood, J.D. and Lawrence, T.L.J., British Society of Animal Production, Occasional Publication No. 17, 1-8

50 Beaumont, J. (1997) *The Consumer Has Spoken*. Key note address, BOCM PAULS Poultry conference, The Belfry, Wishaw

51 Rickard, S. (2001) The future of the animal feed industry in Europe: the effect of CAP and GATT reform. *Alltech's European, Middle Eastern and African lecture tour*, 88-95

52 Roper, P. (1994) (Edit.) *Keyworth 1894-1994. A Century of Change*. Keyworth and District Local History Society, 98-105

53 Taylor, E. (1997) *Kings Caple in Archenfield*. Logaston Press

54 Roper, P. and Wallis, G. (1994) Farming: a changing but continuing occupation.In: *Keyworth 1894-1994*. Edit. Roper, P., Keyworth and District Local History Society, 61-69

55 Hammond, R. (1994) The growth of Keyworth in the twentieth century. In: *Keyworth 1894-1994*. Edit. Roper, P., Keyworth and District Local History Society, 98-105

56 Keyworth Parish Council (2001) *Keyworth Guide*

57 Strak, J. (2000) Planning for change. If it ain't broke get ready to break it. In: *Building Relationships in the Food Chain*. BOCM PAULS conference, The Belfry, Wishaw

58 Cole, J. (1999) *Global 2050. A Basis for Speculation*. Nottingham University Press

59 Cessford, J. (1997) (Ed.) *The Consumer Has Spoken*. BOCM PAULS, The Belfry, Wishaw

60 Cessford, J. (2000) (Ed.) *Building Relationships in the Food Chain*. BOCM PAULS, The Belfry, Wishaw

61 Rixson, D. (2000) *The History of Meat Trading*. Nottingham University Press

62 Pfiffner, A. (1995) Translated by Pulman, D. *Henri Nestlé. From Pharmacist's Assistant to Founder of the World's Largest Food Company*. Nestlé S.A., Vevey

63 Arch, J. (1966) *The Autobiography of Joseph Arch*. Edit. O'Leary, J.G., MacGibbon and Kee, London. First published 1898 as *Joseph Arch: The Story of his Life*.

64 Zuckerman, L. (1998) *The Potato*. Macmillan, London

65 Burrell, M. (1990) *Faith in the Countryside*. Shortened version of the *Archbishops' Commission on Rural Areas*. Arthur Rank Centre. National Agricultural Centre

66 Smith, T. (1857) *The Parish*. H. Sweet, London

67 Loyn, H.R. (1962) *Anglo-Saxon England and the Norman Conquest*. Longman, London

68 Cobbett, W. (1830) *Rural Rides*. 1967 edition with introduction by Woodcock, G., Penguin, Harmondsworth

69 Ellison, W. (1953) *Marginal Land in Britain*. Geoffrey Bless. London

70 Hickman, T. (1977) *The History of the Melton Mowbray Pork Pie*. Sutton Publishing, Stroud

Several organisations' web sites have been consulted.

SOME FURTHER READING ON COUNTRYSIDE AND LAND USE ISSUES

Allanson, P. and Whitby, M. (1996) *The Rural Economy and the British Countryside*. Earthscan Publications

Bowers, J.K. and Cheshire, P. (1983) *Agriculture, the Countryside and Land Use: An Economic Critique*. Methuen

Bunce, R.G.H. and Barr, C.J. (1988) *Rural Information for Forward Planning*. Institute of Terrestrial Ecology. Grange-over-Sands

Cox, G., Lowe, P. and Winter, M. (1986) *Agriculture People and Policies*. Allan and Unwin, London

Dixon, M. (1999) *Principles of Land Law*. Cavendish Publishing, London

Grenville, J. (1999) *Managing the Historic Rural Landscape*. Routledge, London

Hoskins, W.G. (1955) *The Making of the English Landscape*. Penguin, London

Lockhart, D.G. and Ilbery, B. (1987) *The Future of the British Rural Landscape*. Geo Books, Norwich

Miller, F.A. (1991) *Agricultural Policy and the Environment*. Centre for Agricultural Strategy

Mingay, G.E. (1990) *A Social History of the English Countryside*. Routledge, London

Rackham, O. (1986) *The History of the Countryside*. Phoenix, London

Taylor, C. (2000) *Fields in the English Landscape*. Sutton publishing, Stroud

Thirsk, J. (2000) *Rural England: An Illustrated History of the Landscape*. Oxford University Press

Whitby, M.C. and Dawson, P.J. (1990) *Land use for Agriculture, Forestry and Rural Development.* Proceedings, 20th Symposium of the European Association of Agricultural Economists, The University, Newcastle upon Tyne

EPILOGUE

HERACLITUS WAS RIGHT

Those of us who find ourselves towards the end of our careers are sometimes given to feeling the urge to take stock. A sense of the continuity between the historic past, my own past, and the present has always appealed to me. For me, therefore, taking stock includes reviewing the effects of the past on the course of events during my own working life. That accounts for my enthusiasm for the heritage of food, agriculture and the countryside. Some might be inclined to count such an urge as self-indulgence. It is; but I hope that others will agree that it is also useful to use the past to help us to understand the present and to speculate on the future.

Proper historians may wonder why an agricultural scientist would presume to write a history, and my answer is a sense of having lived through some of it. Agricultural technology in Britain has arguably changed more since the 1950s, when I had some hands on experience of it, than it had for the previous 150 years. Household food and food shopping have also changed beyond recognition since the 1940s and 1950s when I was growing up.

We live in times of such rapid change that we need all the clues we can get about the future. To the Greek philosopher Heraclitus change was one of defining characteristics of the world and of human experience. The recent history of food and agriculture might have impressed him as a case in point. The amount of change within my lifetime certainly impresses me.

Let us consider agricultural production first. Threshing machines which Thomas Hardy would have recognised were still in active commercial use in the 1950s, and sheaves of corn were still seen in the fields, though combine harvesters had been in the country since 1928. Hand tools were still used for hedging and ditching and for chopping up root crops to feed to animals. By contrast, cutting hedges is now performed at speed by the kilometre, and mechanical diggers make short work of ditching. Stock feeding is usually mechanical.

There was plenty of hand labour in the fields of the 1950s, of a sort that farmers from the age of George III would have called good practice. Manual hoeing and singling of root crops, and the handling of bales of hay and straw onto trailers with pitch forks, were standard practice; whereas now field machinery is comprehensive and of immense power and speed. Milk churns were still lifted by hand onto lorries for transport to town, whereas bulk tankers now do that job. Often grain and milled animal feed were still manhandled in sacks in the 1950s.

Animal husbandry was generally by methods which would now be called extensive, but by the 1960s came the swing to intensive methods. In the poultry sector this went full circle within the span of my working life. A swing to intensive production, in response to consumer pressure for more and more produce at lower and lower prices, led to scientific and technical questions. From the 1980s came a swing back to extensive systems, also in response to consumer pressure, and with the new swing came new scientific and technical questions.

Farms of the 1950s were small and numerous. A dairy herd of 20 to 25 cows was a viable business for a family farm. By 1999 the average herd size was 73 and more than 8,000 holdings had herds of over 100 cows.[1] Figures given in Chapter 1 show how farms have become much larger but less numerous. Chapter 3 cites similar, but even faster, trends in the poultry sector.

In the 1950s even quite small farms often had several enterprises, whereas specialisation later became necessary. These changes were inevitable and all part of the swing in the driving forces of the market from producer push to consumer pull. In summary the changes have been from small and numerous low tech enterprises producing relatively expensive food, often with government intervention in the market, to large and few high tech enterprises producing cheaper food on freer markets. But we have recently seen a full turn of the circle with the market led revival of low tech systems on smaller units in response to consumer demand for organic, speciality and niche products.

Food supply has gone from the shortages of the years following World War II, with rationing until 1954, to abundance. Consumer concerns have changed from getting enough to eat to avoiding obesity and avoiding excessive fat intake. Obsession with supply has given way to obsession with scares. Food scares and serious questioning and criticism of production practices are now frequent events. With abundance has come choice. Foods which are now every day features of grocery shopping and of eating out, such as quiches and pizzas, were either exotic or scarcely known about in the 1940s and 1950s. When quiches first began to be popular some commentators saw them as slightly posh food.

Feeding children has changed. A 1950s child would have been offered a plate of plain but nourishing food, probably mostly cooked and prepared by the mother from raw ingredients brought home in paper bags, if packaged at all. The 2002 child is offered a range of very attractively packaged prepared foods, often sporting a theme or a cult character image on the packet. The mother may have had little to do with the preparation, beyond opening the packet and worrying about the E numbers.

Shopping and the procurement of the food are different now. In the 1950s much of it came from a corner shop small grocer, whom the shopper knew personally. Most English localities also had a butcher, baker, greengrocer and fish shop. Some of these delivered to the non-car owning shopper. Much of the produce was strictly seasonal; and we have seen how that applied even to staple foods like eggs. Today's motorised shopper visits a supermarket, which is open either around the clock or nearly so, and where irrespective of season a choice of produce from all over the world is offered.[2] Many modern consumers do not know when products are in season in Britain.

With the advent of farm shops, farmer's markets and organic box schemes we are witnessing a small degree of return to older retail methods, but we are also seeing the rise of internet shopping, which was the stuff of science fiction in the 1950s and before. In 1949, in a village school in the New Forest, as 9 year olds we speculated in a class exercise that by 2000 it might become possible to press a button to get our tea on the table. I suppose in the form of internet shopping we were not too far off.

Modern households in Britain nearly all have refrigerators, and most have freezers, so that the regular procurement of perishable food supplies is a less urgent matter. Modern foods include a wider choice of preserved products. Yet modern food shops are open more hours than they were in the 1950s, not less, and for more days in the year, not less. Most supermarkets are open on Sundays, whereas most of the shops of the 1950s were not. In addition there are today's equivalents of the corner shops, in the form of local convenience stores, often open all hours; sandwich bars available on garage forecourts; cafes; restaurants and take-aways. Eating out is now a significant part of life and leisure for many, but it was a rare treat in the 1950s.

Eating out is now an important leisure activity. This tea room is at Middle Farm, Firle, Sussex. Photo: John Pile ©2000 Middle Farm Limited.

The pattern of taking meals has changed in several respects. The cooked breakfast at home has virtually disappeared, except at weekends, though cooked breakfasts are still normal fare at hotels and road houses. By contrast many families of the 1950s started the day with a boiled egg, eggs and bacon or even a complete fry up. Sunday lunch used to be a family occasion of social significance, but it is now rare. Families eating together have generally given way to individuals eating alone, including snacking and fridge raiding, particularly amongst the young. Food has therefore lost, or rather changed, its contributions to social cohesion. Because of the growth in working and business meals and conferences, at which food has strong social functions, it now makes real, but different, social contributions from those which it made 40 or 50 years ago.

Within my lifetime we have witnessed popular attitudes to the scientific approach to food and agriculture change from trust, respect and approval to mistrust, suspicion and sometimes even total rejection. Thus those professing an interest in food and agriculture are apt to be asked if they agree with or approve of the food issues of the day. One answer, perhaps appropriate to many issues, is that offering consumers a choice is a good thing, preferably if it is supported by plenty of clear and reliable information.

Finally, attitudes to the countryside have changed, perhaps because of one particular change in perception. Fifty years ago farms and the countryside were expected to be attractive, but primarily they were associated with food supply and they were therefore expected to be productive. To some extent there has been a reversal of these expectations. There is now a perception that productivity and attractive scenery are separate matters, and if anything farms and the countryside are expected to be primarily attractive and only secondarily productive. This raises new questions about stewardship of the countryside.

None of these observations and contrasts are offered as value judgements. I can see virtues and attractions in the old and the new, and have enjoyed participating in both, but in different ways. They are, however, a declaration of how fascinating it has been to live through such times, and now to have the opportunity to chronicle them.

REFERENCES

1 MAFF (2000) *Agriculture in the United Kingdom 1999*. The Stationery Office

2 Seth, A. and Randall, G. (1999) *The grocers*. Kogan Page, London

INDEX

References to the Chronology are generally omitted from the Index, except for topics receiving little or no mention in the text.

Acre, 8
ADAS Gleadthorpe, 94
ADAS, 47
Added value, 3, 33
Adulteration of Foods Act, 112
Advisory services, 40
Agricultural holdings, numbers of, 19
Agricultural Marketing Act, 24
Agricultural sciences, contribution of, 28
Agricultural Society of England, 27
Agricultural Training Board (ATB), 43
Allotments, 24, 88
Amenity, 144
Animal welfare, 113, 127
Anti-Corn Laws League, 20
Appert, 36
Apples, 74
Arbuthnot, The Hon.Mrs., 84
Arcadia, 112
Arch, Joseph, 2, 23
Aristotle, 30
Arourae, 7
Art, animals in, 63
Art, textbook illustrations as, 6, 7
ASDA, 120
Attitudes to food, 113
Atwater and Bryant's food analysis tables, xvi
Aurochs, 61
Austen, Jane, 110, 123
Aviaries, 90
Axholme, Isle of, 8
Baby food, xiv
Bailey, F., 129
Bakewell, Robert, 13, 32, 64, 87, 125
Bantams, 82
Battery cages, and alternatives to, 89
Baxter's *Library of Agriculture*, 60, 61
Beeton, Mrs., xiv
Bell, Patrick, 15
Biodynamic movement, xvii
Bioenergetics, 30
Biomass, 74
Bird's custard, 38
Black Death, x
Blaxter, Sir Kenneth, 31, 33
Blickling, Norfolk, 14
Board of Agriculture, 24, 27, 41
BOCM PAULS, 33
Bottling, 36
Box schemes, 153
Boyd Orr's report, xviii
Brahman, 61
Brambell Report, 99
Bramley apple, 75
Brands and branding, 120
Brassica, 77
Bread, 72
  barley, 72, 73
  price of, 19
  sliced, xvii
  wheaten, 72, 73
Breakfast, cooked, decline of, 99, 154
Brett, Mr., 12
British Egg Marketing Board, 99

British Poultry Council, 99
British Poultry Science, 95
Brody, Samuel, 30, 31
Broilers
  introduction of, 95
  performance of, 101
Burnham Market, Norfolk, 24
Butter making, 34

Cabbage, 77
Cake mixes, xvii
Calorie, 30
Calorimetric measurements, 31
Candia, 36
Canning, 36
Capel Manor College, 47
Carrot, 77
Cat, domestication of, 69
Cattle plague, 35
Cauldrons, 55
Cave paintings, 58
Central Statistical Office, 3, 33
Cereals, 67
Ceres, 69
Cheese, 34, 63
  Cheddar, 34
  press, domestic, 35
  ration, World War II, 25
  Stilton, 44
  variety of, 128
Chips, potato, 75
Chocolate, 120
Chocolate, milk, xv
Chorleywood bread process, xx
Church of England, 141
Churches, 141
Civilisation, birth of, 67
Clover, 9, 77
Coates's Shorthorn herdbook, 64
Cobbett, William, 2, 6, 23, 76, 77, 115
Cod liver oil, 25
Coke of Holkham, Thomas, 9, 12, 40
Coles, Dr. R., 95
Colleges, county council, 43
Colleges, history of, 43
Colling brothers, 14, 64, 65
Columella, 5, 6, 74, 84, 87
Combine harvester, 15
Common Agricultural Policy (CAP), xx
Commuting, 125, 137
Computers , early industrial application of, 95
Coneys, 61
Consumers
  attitudes to food and agriculture, 113, 129
  choice for, 124
  expectations of, 122, 125
  expenditure by, 3
  expenditure on food, 3
  power of, 33, 125
Consumption trends
  eggs, 114
  meat, 114
  milk, 114
  vegetables, 115
Controlled environment housing, 94, 97
Convenience foods, 128
Cook shops, London, 37
Cookery, 18th century, 1
Cooking, 55
Co-operative retailing, 36
Corn laws, 20, 23, 111
Corn Protection Act, 23
Cottage economies, 60
Cottagers, 18
Council for the Protection of Rural England (CPRE), 139
Country Landowners Association (CLA), 139
Countryside
  expectations of, 131, 137
  residential, 135, 144
  stewardship of, 144
  uses of, 144
  used for sports and leisure, 134
Country sports, 134, 144
Countryside Agency, 132, 139

Countryside Alliance, 142
Countryside Commission, 139
Cow keepers, urban, 37
Crisps, potato, xiv
Cruciferae, 77
Crystal Palace, 88
Curry Report, xxii
Custard, Bird's, 38

Dairy work, colour section
Daubeny, Prof. Charles, 41
Davy, Sir Humphrey, 27
Decker's *Cheese Making*, xvi
Deduction, 30, 113
Deductive nutrition, 30, 113
DEFRA (Department for Environment, Food and Rural Affairs), 2, 132, 139
Degrees, 43
Demeter, 69
Department for Environment, Food and Rural Affairs (DEFRA), 2, 132, 139
Dietary standards, Edward Smith's, xiv
Dig For Victory, 25
Digestive biscuits, xiii
Diocletian, 19
Diplomas, 43
Directives, 127
Dishley, Loughborough, 64
Diversification, 136
DNA fingerprinting, 68
Domesday records, 110
Domestication, 57, 58
  bees, 63
  cats, 69
  cattle, 57, 61
  cereals and grasses, 67, 69
  chickens, 82
  ducks, 85
  fruits and vegetables, 74
  geese, 85
  goats, 60
  legumes, 74
  pigs, 60
  potatoes, ix, x, xi
  rabbits, 61
  sheep, 60
  turkeys, 84
Donkin, 36
Dorking, 84, 87
Dovecotes, 92
Doves, 86, 92
Drainage, 135
Drinking chocolate, xi
Drovers, of turkeys and geese, 85
Duhamel, 13
Dunstan, M.J.R., 43
Durham Ox, 14, 65

Easton College, 47
Eating out, 3, 34, 37, 153
Eating, social and cultural significance of, 123
EC, 127
EC Agenda 2000, 27
Education and training, agricultural, 40
Edinburgh School of Agriculture, 47
Eggs
  consumption trends, 114
  lighting for production of, 94
  production holding size, 95
  sales of, 99
  seasonality of supply of, 93
  systems of production of, 95, 99
  yields of, 101
Eggs Authority, 99
Eglantine Vineyard, 76
Egypt, ancient, 7, 63, 122
Einkorn, 68
Electronic offices, 137
E-mail, 137
Enclosure, 8, 116, 117
Energy
  agricultural and industrial use of, 143
  crisis, 97, 143
  metabolisable, 32
  net, 32
Engineering, agricultural, 15
English Heritage, 139

English Nature, 139
Epworth, 8
Ernle, Lord, 7, 116
Ethnic foods, 39
Euro PA, 33
Experimental plots, 28

Fairs, goose, 86
Fallow, 8, 9
Farm business size, 152
Farm business, 19, 95
Farm shops, 2, 153, colour section
Farmers' markets, 2, 120, 121, 153, colour section
Fast food, 37
Feeding systems for farm animals, early, 32
Feeding trials, 32
Feeds, animal, 32
Ferguson, Harry, 18
Fermentation, 35
Fire, use of, 55
Firle, Sussex, 128, 153, colour section
First aid, 43
Fish and chip shop, mobile, 39
Fish and chip shops, 37
Fish fingers, xix
Flatulence, xiii
Fodder, winter, 63
Food
  analysis, tables of, xvi, 29, 32, 33
  attributes of, 130
  convenience, 128
  emotional values of, 130
  household, 2, 3, 34
  manufacture, 55, 122
  preservation of, ix, 36
  rationing, 25, 26, 27, 143
  retailing, 122, 123, 124, 125
  safety, 114, 124
  scares and fashions, 122, 123, 124, 125
  security, 145
  social significance of, 123
  speciality and regional, 128
  spending on, 2, 3
  surveys, 129
  tools for preparation of, ix
  trade, medieval regulation of, 19, 37
Food Safety Act, 112, 125
Food Standards Agency, 125
Food Technology, xix
Food trade, pressures and issues, 133
Foot and mouth disease, xx, xxii
Forecasting, hazards of, 142
Forecasts, 145
Forestry Commission, 131
Forestry, 131
Fox, Charles James, 18
Frazer, Mrs, 1
Free trade, 19
Freedom Food, 100
Freezers, 153
Fruit and vegetables, 2, 75, 113
Future, issues for, 144

Galen, 122
Game, small, 55
Garden stuff, 2
Gas cookers, xvi
Geese, manure of, 9
General Agreement on Tariffs and Trade (GATT), xxi
Genetics, 77
Geometry, 69
George III, 151
Germ theory of food spoilage, xiv
Gilbert, 23
Glass house crops, xv
Gleadthorpe, ADAS, 94
Globalisation, 123
GM foods, 125
Googe, Barnaby, 15, 76
Grafting, 74
Graminae, 77
Grasses, 67

Great Western Railway Company, 19
Greeks, ancient
  Arcadia, 112
  art, 63
  diet, 122
  Egypt, in, 7
  farming, 125
  fruit and vegetables, 74
  mathematics, 122
  nutrition, 30
  soil fertility, 9
Green sickness, 65

Hadlow College, 47
Haldane Committee, xvii
Hammond, Sir John, 94
Harding, Joseph, 34
Hardy, Thomas, 2, 116, 151
Harper Adams University College, 43, 46, 94
Harrison, Ruth, 99
Hartpury College, 47
Harvest festivals, 110, 111
Harvest jug, 111
Harvesting, 71, 72
Health food, 37
Hedges, 116
Hekat, 32, 122
Heraclitus, 142
Herbae inutiles, 13
Herbs, introduction to Britain of, x
Herd size, dairy, 152
Herefordshire, 136
Herodotus, x
Hesiod, 5
Highland clearances, 119
Histories, local, 136
Hobart, Sir Henry, 14
Holdings, agricultural, numbers of, 19
Holidays, farm, 136
Holkham Hall, 40
Holkham sheep shearings, 40
Holkham, Norfolk, 9
Home and Colonial, 37
Home working in the countryside, 137
Honey, 63
Horses, working, 15, 68
Household food, spending on, 2, 3
Household income, 127
HP Sauce, 37
Human nutrition, 33
Humours, theory of, 30, 122
Hunter gathering, 55, 118
Hybrid poultry, 83
Hybridisation, between wild and domestic, 56
Hyett, 30

Ice houses, 94
IGD, 33
Incubators, 90
Indoor pigs and poultry, 81
Induction, 30, 113
Inductive nutrition, 113
Institute of Grocery Distribution (IGD), 33
Integrated Pollution Prevention and Control (IPPC), 95
Intensive livestock farming, 81
Internet shopping, 153

James, W.P.T., 33
Joule (J), 30
Joule, James, 30
Journal of Agricultural Research, xvii
Journal of Agricultural Science, xvi
Journal of the Science of Food and Agriculture, xix
Jungle fowl, 83

Kellner, Oscar, 32
Keyworth, Nottinghamshire, 137
Kilner jars, xiv
Kings Caple, Herefordshire, 136
Kingston, Nottinghamshire, 44
Kleiber, M., 31

Laboratories, agricultural, 31
Land Settlement (Facilities) Act, 24
Land Settlement Association, 24
Lavoisier, xii
Laxton, Nottinghamshire, 8, 9, 10, 11
Laying trials, county, 43
Least cost feed formulation, 95
Legumes, *Leguminosae*, 74, 77
Leisure, 134, 144
Liebig, 29, 30, 32
Linseed cake crusher, 63, 64
Lipton's, 37
Lloyd George, 48
Local histories, 136
London, medieval, 1
Longhorn cattle, 13
Lyon's Corner Houses, xvii

MAFF (Ministry of Agriculture, Fisheries and Food), 2, 27, 33, 41, 132, 139
MAFF bulletins, 97
Malthus, 143
Mangold-wurzels, 32, 77
Manhattan, 118
Manure, poultry, 9
Manuring, 9
Margarine, 36
Marginal lands, 112, 131, 135
Marmalade, x
Mathematics and mensuration, 69
McCance and Widdowson's *The Composition of Foods*, 29, 32, 33
McCormick, Cyrus, 15
Meat
  aspirational, 112
  consumption trends, 114
Medieval towns, regulation of food trade of, 19, 37
Meikle's threshing machine, xii
Melton Mowbray farmers' market, 121
Melton Mowbray pork pies, 104
Mendel, Gregor, 77
Menus, meat and two vegetables, 56
Merryweather, Henry, 75
Michaelmas goose, 86
Middle Farm, Firle, 128, 153, colour section
Midland Dairy Institute, 42
Milk and Dairies Act, xvii
Milk consumption trends, 114
Milk Marketing Board, 24, 35
Milking, hand, 45
Ministry of Agriculture, Fisheries and Food (MAFF), 2, 27, 33, 41, 132, 139
Ministry of Food, 25
Monogamy, 118
Mouflon, 60
Mulder, xiii
Muscovy duck, 85

NAAS (National Agricultural Advisory Service), 47
Napoleon, 36
National Agricultural Advisory Service (NAAS), 47
National Farmers' Union (NFU), 27, 139
National Food Survey, 3
National Institute of Poultry Husbandry, xviii
National Parks, 132
National Trust, 139
Natural habitats, 144
Natural History of Selborne, The, 73
Nature reserves, 139, 144
Naturescape, 132, colour section
Neolithic, 67
Nestlé, 123
Niche markets, 100, 127
Norfolk four course rotation, 9
Nottingham University Press, 137
Nottingham, University of, 43, 44
Nottinghamshire District Agricultural Schools, 43
Nutmeg, 118
Nutrition, human, 33

Oats, 70

Office for National Statistics, 127
Orchards, ix, 74
Organic products, 100, 128
Organisations, statutory, charitable, non-government, 138
Otley College, 47
Ovens,
  Celtic ground, 55
  hole and stone, ix
Ox teams, 15, 119

π, 122
Palaeolithic, 55, 118
Palladius, 86
Paracelcus, 122
Parishes, 141
Parkinson, Frank, Agricultural Trust, 47
Partnerships, 142
Pasteur, xiv
Pasteurisation, 36
Peas, 74, colour section
Pefsu, 32
Pershore College of Horticulture, 47
Peters, Matthew, 6
Pheasants, 82
Pigeons, 86, 92
Pigs, 60, 81, 102
  domestication of, 60
  feed conversion of, 97
  indoor, 103
  market changes, 102
  modern breeds, 103
  urban, 37
Playfair, 32
Plough, mould board, social effect of, 119
Poor laws, 23
Population, proportion in agriculture, 3, 4, 143
Population, world, 142
Pork pies, 38, 102, 104, 129
Pork pies, Melton Mowbray, 104
Pot herbs, 2
Potato blight, 119
Potato Marketing Board, 24
Potatoes, 74, 75, 76
Poulterer's Livery Company, 87
Poultry, 81
  business sizes, 96
  controlled environment, 94
  crammer, 90
  domestication of, 82
  feed conversion of, 97
  free range, 89, 92, 93, 100
  holdings with, 96
  housing, 96
  hybrids, 83, 86
  instructors, peripatetic, 41, 88
  manure, 9
  meat production, 101
  niche markets, 100
  sales and marketing of, 98
  ventilation of, 97
Poultry Show, Poultry Fair, 87
Prediction, hazards of, 142
Preservation of food, 36
Protein, xiii
Prothero, R., (Lord Ernle) 7, 116

Railway Company, Great Western, 19
Railways, effect of, 23
Ramsden's , Harry, xviii
Range arks, 81, 92, 93
Rapeseed, colour section
Rare Breeds Survival Trust, 65, 88
RASE (Royal Agricultural Society of England), 7, 15, 27, 28, 29, 41
Ration books, 26
Rationing, food, 25, 26, 27, 143
Reaper, 15, 16
Refrigeration, 36
Refrigerators, domestic, 153
Regional and traditional foods, 36, 38
  variety of, 128
Regulations, 127
Research, agricultural, 40
  Agricultural and Food Research Council, 48

Agricultural Research Council, 48
Biotechnology and Biological Sciences Research Council, 48
Residential use of countryside, 135, 144
Restaurant, first, xii
Rice, 69
Richmond, Duke of, 20, 23
Ridge and furrow, 8
Robinson, Leonard, 89
Rochdale, Equitable Pioneers of, 36, 120
Romans
art, 63
chickens, 84
diet, 122
fruit and vegetables, 74
herbs, x
poultry industry, 86
salt, 36
villas, 8
wheat price, 19
Root cutter, 12
Rotations, crop, 9
Rothamsted, 28
Round bales, 117
Rowett Research Institute, 33
Royal Agricultural College, 42, 47
Royal Agricultural Society of England (RASE), 7, 15, 27, 28, 29, 41
Royal Commission 1897, 24
Royal Show, 15, 35
Royal Society for the Prevention of Cruelty to Animals (RSPCA), 100, 141
Royal Society for the Protection of Birds, 141
Run, island of, 118
Rural employment, 20,137
effects of electronic offices and home working, 137
future of, 144
Ryegrass, 67

Sacrewell Farm and Country Centre, 33, 88
Sacrificial animals, 58
Safeway, 120
Sainsbury, J., 37, 120
Salmonella, 100
Salt licks, 57
Salt, 36
Sampling techniques, 28
Sandwich, Earl of, 37
Sandwiches, 37
Sawrey, Lake District, colour section
Scavenging, ix
Scientific Poultry Breeders Association, 88
Scythe, x
Seal Hayne, 42, 47
Self service stores, xvii, 37
Shopping experience, 130
Sittingbourne College, 47
Small Holdings Act, 24
Smallholdings and Allotments Act, 24
Smallholdings, 24
Smithfield, 85
Social background, 18
Society, 110
Soil Association, xix, 128
Soil fertility, 9
Solonacae, 77
Soup kitchens, xvii
Soya beans, 74
Sparsholt College, 47
Speenhamland System, 23
Spencer, Earl, 32
Spice wars, 118
Spices, 2, 4
Sports, country, 134, 144
Stapledon, Sir George, 131, 135, 145
Starch Equivalent, 32
Statistical analysis, tables for, 28
Statistics, 28
Steam ploughing, 17
Steiner, Rudolph, xvii
Strip cultivation, 8
Sugar, 36,122
beet, 77
cane, 74

refinery, 36
Sunday lunch, 154
Supermarkets, 120, 153
Sutton Bonington, 42, 46
Swans, 87
Swede, Swedish turnip, 77
Swill mixer, 89

Tannahill, R., 7
Tea shops, 37
Tea, xi
Temperton Report, 98
Tesco, 120, 123, colour section
Thales of Miletus, x
Three field system, 8, 9, 10, 11
Threshing, 116, colour section
Thrumpton, Nottinghamshire, 137
Tolpuddle Martyrs, 23
Tomatoes, xv
Tools, stone age, 56, 57, colour section
Tourism, 144
Toussaint-Samat, M., 7
Townshend, Turnip, 9
Traceability, 2
Tractors, 18, 155, 156, colour section
Tractors, numbers of, 15
Traditional and regional foods, 38
Travelling instructors, 43
Tull, Jethro, 13, 14, 15
Turnips, 9, 12, 32, 77
TV cooks, 40

UK Poultry Research Liaison Group, 99
UK Vineyards Association, 75
Undernourishment, 25
University of Nottingham, 42, 44
Utility Poultry Club, 101
Uttoxeter, Staffordshire, 41

Vegetables, 2, 74, 75, 113
Vegetables, consumption trends, 115
Vegetarianism, 55, 65
Villages, changes in, 135
Vinegar, 37
Vines, 75, 76
Visionaries, 145
Vitamin B_{12}, 72
Vitamin C, 75
Vitamins in animal nutrition, 95

War Agricultural Executive Committees, 25
Warreners, 61
Water glass, 94
Water mill, 73
Watt (W), 30
Weed control, 13
Weston, Sir Richard, 14
Wheat, 69, 70
Whisky Money, the, 41
White, Gilbert, 73
Wild flower farm, colour section
Wild flowers, 132, colour section
Willoughby on the Wolds, Nottinghamshire, 8
Wilson, Prof. P., 47
Women's Institute, 25, 139
WI and the war effort, 140
Women's Institute markets, 120, 128
Women's Land Army, 16, 21, 22, 91
Women's Mission to French farms, 20
World's Poultry Science Association, 82, 88
Woodland Trust, 141
Wordsworth, William, 18, 112, 119
World War I, 20, 25, 88, 112, 140
World War II, 22, 25, 33, 38, 88, 89, 91, 112, 140, 152
Wright of Derby, Joseph, 29
Wye College, 47

Yields, improvements in animal, 65, 101
Youatt, W., 6
Young, Arthur, 5, 27

Zebu, 61